International Cryogenics Monograph Series

The International Cryogenics Monograph Series was established in the early 1960s to present an opportunity for active researchers in various areas associated with cryogenic engineering to cover their area of expertise by thoroughly covering its past development and its present status. These high level reviews assist young researchers to initiate research programs of their own in these key areas of cryogenic engineering without an extensive search of literature.

More information about this series at http://www.springer.com/series/6086

Milind D. Atrey
Editor

Cryocoolers

Theory and Applications

Springer

Editor
Milind D. Atrey
Refrigeration and Cryogenics Laboratory
Department of Mechanical Engineering
IIT Bombay
Mumbai, India

ISSN 0538-7051 ISSN 2199-3084 (electronic)
International Cryogenics Monograph Series
ISBN 978-3-030-11306-3 ISBN 978-3-030-11307-0 (eBook)
https://doi.org/10.1007/978-3-030-11307-0

This Springer imprint is published by the registered company Springer Nature Switzerland AG.
The registered company address is: Gewerbestrasse 11, 6330 Cham, Switzerland

Preface

Cryocoolers are important devices in various applications such as MRI machines and cryopumps. They are reliable, simple to use, and are cost-effective. The use of cryocoolers has increased manifold since 1980 and most of the equipment which was earlier cooled by liquid nitrogen and liquid helium is now cooled by cryocoolers like G-M and pulse tube cryocoolers. Even the dilution refrigerators, which produce temperature below 1K, have witnessed the use of cryocoolers to reach up to 2–3K first for precooling purpose.

In view of the increasing requirement of cryocoolers in MRI machines, small capacity liquefiers, dilution refrigerators, fundamental physics experiments, etc., the cryocooler technology has become efficient and innovative. In view of this, a need was felt to write a book which would describe different types of cryocoolers, their working, design considerations, and various applications. I have been researching in this field for more than 30 years and have worked on almost all the types of cryocoolers. My experience of working with University of Giessen, Germany, Oxford Instruments Superconductivity, UK, and now at IIT Bombay, India, exposed me to different types of cryocoolers. An attempt is made in this book to showcase the use of different types of cryocoolers for various applications, with many well-known researchers in this area contributing to this book. I believe this compilation will be of great value and relevance to the students and researchers in the field of Applied Cryogenics and will serve as a ready reference in their work. I extended my heartfelt thanks to all the authors who contributed to this book and equally heartfelt apologies for the delay in the publication.

The involvement of so many authors creates complex problems of coordination, editing, and timelines. During the editing of this book, I faced several such problems to meet the timeline, but ultimately managed to complete the book. Several people have helped me in editing this book: namely my graduate student, Sana Syed and Prabhakar who is an engineer in the Cryogenics lab at IIT Bombay. Sana has contributed immensely in the matter of getting approvals for the various photos for my chapter, while Prabhakar helped me with the formatting of the chapters and incorporating all the corrections. Lastly, I remain grateful to my wife, Dr. Mukta, who offered her expertise in language and editing. She went through each chapter meticulously, finding typos and linguistic aberrations. She also communicated with each author to get the chapters and the final go-ahead from each of them. I thank all of them sincerely for their assistance.

Mumbai, India Milind D. Atrey

Contents

1 **Cryocooler Technology: The Path to Invisible and Reliable Cryogenics** 1
Milind D. Atrey

2 **Joule Thomson Cryocoolers and Cryoablation** 47
Harrison M. Skye and John M. Pfotenhauer

3 **Mixed Refrigerant Joule–Thomson Cryocooler for Cooling High-Temperature Superconductor Cables** 65
Sangkwon Jeong, Jisung Lee, and Cheonkyu Lee

4 **Cryocooler Applications at Neutron Scattering Facilities** 83
Oleg Kirichek

5 **Integration of Adiabatic Demagnetization Refrigerators with Spaceflight Cryocoolers** 99
Peter J. Shirron and Michael J. DiPirro

6 **Cryocoolers for Superconducting Generators** 121
Wolfgang Stautner

7 **Cryocoolers for Laboratory Research** 155
Ravi Bains

8 **The Diaphragm Pressure Wave Generator** 183
Alan Caughley

9 **Cryocoolers for Healthcare** 191
Wolfgang Stautner

10 **Cryocoolers for Helium Liquefaction** 205
Stefano Spagna, Javier Sesé, Miguel Gabal, and Conrado Rillo

About the Editor

Milind D. Atrey is currently Dean (Research and Development) and Chair Professor at IIT Bombay, India. A Mechanical Engineering graduate of VNIT Nagpur, he obtained his Doctorate from IIT Bombay in 1991. He works in the area of Refrigeration and Cryogenic Engineering in the Department of Mechanical Engineering, IIT Bombay.

Prof. Atrey's experience in the area of Cryogenic Engineering is a unique blend of corporate research, work in government research laboratories and academia. After a brief stint with the Tata Research Development and Design Centre, Pune, after completing his doctorate, he worked on the three-stage Stirling cryocooler and 4 K G-M coolers for his post-doctoral research at the University of Giessen, Germany. Subsequently, he worked at the Raja Ramanna Centre for Advanced Technology (RRCAT, Department of Atomic Energy, Govt. of India) on a project aimed at the indigenization of a cryogenic technology. Prof. Atrey returned to corporate research 5 years later, joining the Technical Development group at the Oxford Instruments Superconductivity, Oxford, UK. His body of work includes NMR/MRI cryogenic technology, cold probe technology for NMR, and the development of 4 K pulse tube cryocoolers. The cold probe technology developed by his group proved to be path breaking for NMR researchers. The group also demonstrated the world first Helium recondensing technology on NMR superconducting magnet.

In the last 15 years at IIT Bombay, Prof. Atrey has developed various pulse tube cryocoolers, mixed refrigerant Joule–Thomson cryocoolers, thermoacoustic coolers, etc. His recent research includes cryogenic two phase flow, pulsating heat pipe, heat exchangers, cryosurgery, etc.

Prof. Atrey is a Fellow of the Institute of Mechanical Engineers (FIMechE) as well as the Indian Cryogenics Council. He takes over as President, Indian Cryogenics Council in April 2020.

Chapter 1
Cryocooler Technology: The Path to Invisible and Reliable Cryogenics

Milind D. Atrey

Abstract Cryocooler technology has come a long way since the 1950s, following the development of Stirling cryocoolers for air and hydrogen liquefaction by the Philips company in the Netherlands. This chapter touches upon both the introductory and advanced aspects of cryocooler technology. It starts with a detailed introduction on cryocoolers and the need for these devices for industrial applications for cryogenic temperature refrigeration. Further, it describes the working for each type of cryocooler and then goes on to highlight the current trends in cryocooler technology in the past decade. It also lists out the applications and challenges faced by researchers and the cryocooler industry for further development in this area. The motivation for writing this chapter is to provide both novice and veteran readers exploring cryocooler technology with all the relevant information about cryocooler development and direct them towards exploring further in the exciting field of cryogenic refrigeration devices.

1.1 Introduction

Cryocoolers or cryo-refrigerators are devices used to produce refrigeration at cryogenic temperatures, which is normally below 123 K [1]. Such low temperatures have many applications, some of which are given below.

- Gas liquefaction and separation
- Cooling of superconducting devices
- Cryopumps for clean vacuum
- Tissue ablation (cryosurgery)
- Preservation of biological material and food
- Cooling of electronic devices with low thermal noise

M. D. Atrey (✉)
Refrigeration and Cryogenics Laboratory, Department of Mechanical Engineering, IIT Bombay, Mumbai, India
e-mail: matrey@iitb.ac.in

M. D. Atrey (ed.), *Cryocoolers*, International Cryogenics Monograph Series,
https://doi.org/10.1007/978-3-030-11307-0_1

Most cryocoolers work in a closed cycle and use helium (He) as the working fluid. This is due to the fact that it has a low boiling point (4.2 K) and an inert nature. However, some cryocoolers use hydrogen (H_2), neon (Ne) or gas mixtures. In today's scenario, cryogen reserves are depleting at an alarming rate leading to an increase in cryogen prices, especially for helium gas. Cryogenic applications such as cooling of superconducting magnets utilize large amount of cryogens like liquid nitrogen (LN_2), liquid helium (LHe), etc. These cryogens need to be replenished at regular intervals to maintain the required low temperature. A popular newspaper article spells out the looming urgencies, which the world will face, due to rapid depletion of helium reserves in the years to come [2].

Figure 1.1 shows the approximate current helium prices in selected places in the world, based on estimates and data from suppliers [3]. With depleting cryogen reserves, cryocoolers that produce cooling effect at low temperatures in a closed cycle manner have gained importance in a multitude of applications due to the inherent advantages.

On the other hand, some cryocoolers where cool-down time is very important, like Joule–Thomson cryocooler, are open cycle systems and may require very high pressure (100–400 bar) for particular applications. Improved reliability and advancement in cryocooler technology in the past two decades have led to an increase in the use of cryocoolers for space, industrial and medical applications [4]. This chapter reviews the salient features of the available cryocooler technology and discusses a few analysis methods for cryocooler and highlights the latest developments in this field.

Fig. 1.1 Helium prices around the world in select regions [3]

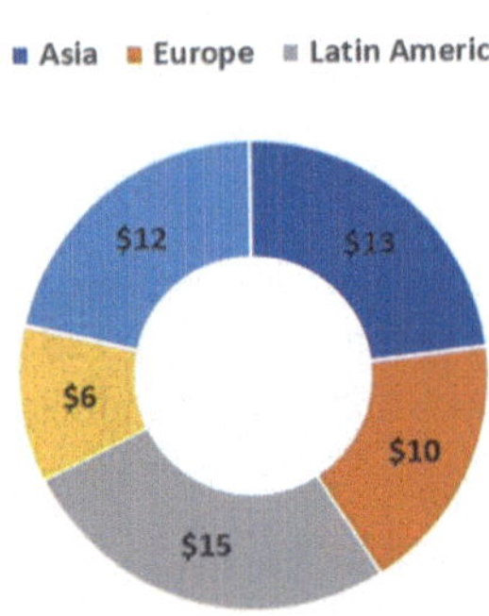

US	$3/litre and $9/litre in case of emergency supply
UK	$6/litre

Current prices of Nitrogen and Helium in India		
Cryogen	**Cost**	**Source**
Helium gas	₹ 1500 / m^3	Linde India Limited
Liquid Helium	₹ 1300 / litre	
Nitrogen Gas	₹ 150 / m^3	
Liquid Nitrogen	₹ 20 / litre	

INR 68.73 = 1 USD

1.2 Classification of Cryocoolers

Cryogenic refrigeration can be achieved by using a cryocooler or by use of cryogen to maintain low temperatures. Figure 1.2 gives various techniques by which cryogenic refrigeration can be achieved using different types of cryocoolers [5].

In open cycle systems, cryogenic refrigeration is obtained by using an open cycle Joule–Thomson (J-T) cooler, which uses gas stored at high pressure and maintained at ambient temperature. The stored gas is released to the atmosphere after expansion. Most applications, however, use stored cryogens such as liquid nitrogen (LN_2) and liquid helium (LHe) to maintain cryogenic temperatures. These stored cryogens need to be replenished at regular intervals. They could be in solid, liquid, or supercritical form as per the requirements of the application. However, for getting stored cryogens, large-scale refrigeration plants may be used for the liquefaction where different cycles of refrigeration for liquefaction may be used.

In closed cycle systems, either dynamic (existence of moving component) or static (non-moving component) devices are used to attain cryogenic temperatures. Dynamic devices may include compressor piston arrangement with or without a moving expander. Such cryocoolers can be further classified based on the type of

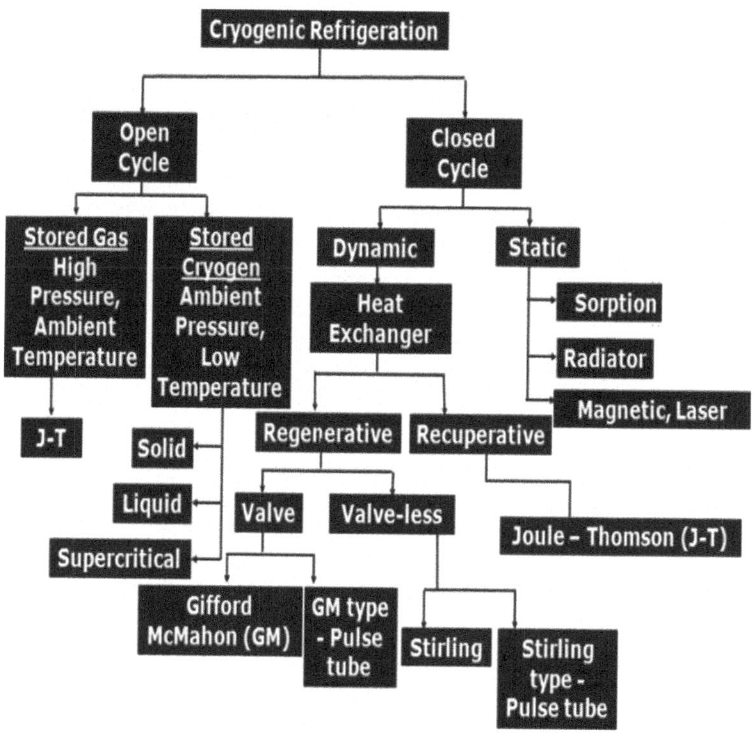

Fig. 1.2 Cryogenic refrigeration [5]

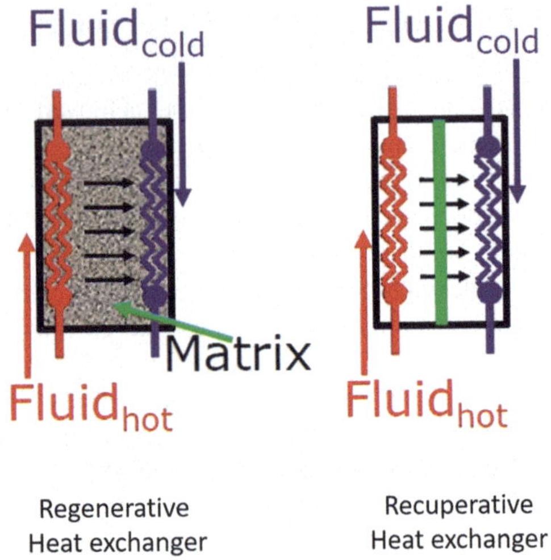

Fig. 1.3 Types of heat exchangers for cryocoolers [5]

heat exchanger used in the system, i.e. recuperative type or a regenerated type. Figure 1.3 shows a schematic diagram of both types of heat exchangers.

In a recuperative heat exchanger, the hot and cold fluid streams flow simultaneously through their respective channels, thereby transferring heat from one stream to another. The Joule–Thomson (J-T) cryocooler is a good example of a recuperative cryocooler. Regenerative heat exchangers are characterized by the presence of a regenerator matrix, which serves as an intermediate heat exchange medium between the warm and cold fluids. The flow through this matrix is periodic in nature, alternating between the warm and cold fluids, and thus providing a cooling effect at the cold end of the regenerator. The lowest temperature that can be obtained using such cryocoolers depends on the heat capacity of the regenerator matrix material at lower temperatures. Regenerative cryocoolers can be of two types, depending on the presence of valve between the compressor and the cold head. The cold head comprises a regenerative heat exchanger and an expander. The valved systems are called as Gifford McMahon (G-M) type cryocoolers, while systems without valves are known as Stirling-type cryocoolers. This classification is explained in detail in the subsequent sections.

Static cryocoolers, as mentioned earlier, have no moving parts. One example of such static cryocoolers is a sorption compressor-based J-T cryocooler [6]. Radiative cryocooler is another example, generally used in space applications. Such cryocoolers make use of high reflectivity surfaces, which provide thermal shielding to the cooled unit from space. These surfaces also reflect the solar radiation falling upon the unit [7]. Other examples are systems which use magnetic or laser cooling mechanisms. Adiabatic demagnetization refrigerator (ADR), which is based on materials exhibiting a strong magnetocaloric effect, is an example of magnetic cooling. In laser cooling, the force of a laser beam on an object nearly halts the motion of its atoms. This reduces their stored energy and hence results in lowering of its temperature. Static cryocoolers are preferred over dynamic ones where the application demands a vibration-free atmosphere and very high reliability. However, they may not be cost-effective for other applications.

1.3 Types of Cryocoolers

This chapter deals with the cryocoolers that are generally used in industry. These are usually of the dynamic class with a moving part in the compressor. Static cryocoolers, which form less than 2% of the cryocoolers used worldwide, are not discussed here. In this section, first the working of regenerative heat exchanger-based cryocoolers is explained, while those with recuperative heat exchangers are dealt with later. The regenerative cryocoolers explained are the Stirling, Gifford-McMahon (G-M) and pulse tube cryocoolers, while the recuperative cryocooler is the J-T cryocooler.

1.3.1 Stirling Cryocoolers

Stirling cryocoolers were the first developed cryocoolers in 1950s and are by far the most efficient of all types of cryocoolers till date. All other regenerative cryocoolers may be considered as the modifications of the Stirling cryocooler. The reverse Stirling cycle, based on which the Stirling cryocooler works, was first developed as an engine cycle by Robert Stirling in 1815. Almost 100 years later, commercial cryocoolers based on Stirling cycle were first developed in the 1950s by the Philips Company in the Netherlands for air and hydrogen liquefaction [8]. Since 1960s, miniature Stirling cryocoolers are being used as cooling devices of infrared (IR) sensors for communication purpose. For these applications, the required cooling effect is around 1–2 W at 80 K. In the 1970s and 1990s, efficiency and reliability of these cryocoolers improved due to the introduction of flexure bearings and gas bearing support systems in the compressor. Stirling cryocoolers are now heavily used in night vision surveillance equipment. Further research is being carried out to miniaturize the cryocooler and attain lowest possible temperatures for military and space applications.

Figure 1.4 shows the schematic diagram and working principle of Stirling cryocooler [5].

As seen in Fig. 1.4, a Stirling cryocooler consists of a compressor, a regenerative heat exchanger called regenerator, and an expander piston/displacer housed in an expansion chamber. As mentioned earlier, the Stirling cryocoolers are valve-less units, and hence, the expander moves at the same frequency as that of the compressor piston with a fixed phase angle between their motions. This phase difference between the compressor piston and the expander-displacer is vital for an

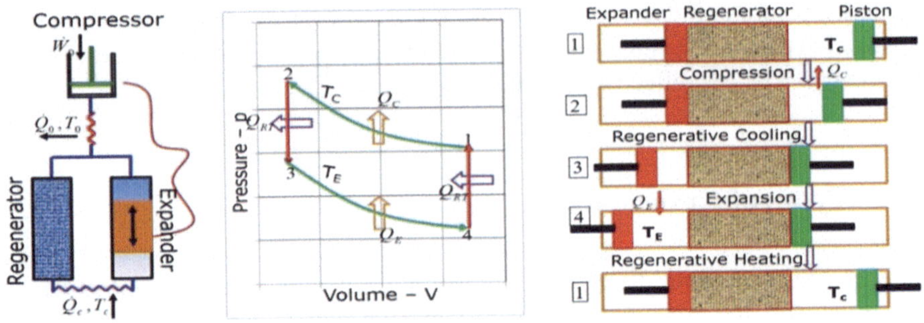

(a) (b) (c)

Fig. 1.4 (**a**) Schematic of Stirling cryocooler, (**b**) Stirling cycle, (**c**) working principle of Stirling cryocooler

efficient operation of the Stirling cryocooler. In most of the miniature cryocoolers, the hollow displacer houses the regenerator matrix.

As seen from Fig. 1.4, an ideal Stirling cycle consists of four processes: isothermal compression (1–2) at T_c, isochoric heat rejection to regenerator matrix (2–3), isothermal expansion (3–4) and isochoric heat absorption from regenerator matrix (4–1). The COP of an ideal Stirling cycle is equal to the COP of a Carnot cycle.

Figure 1.4 shows the working principle of a Stirling cryocooler. At state 1, the piston is at its bottom dead centre (BDC), and the expander is at its top dead centre (TDC). The process of compression takes place when the compressor piston starts moving towards its TDC (1–2), during which heat of compression, Q_c is released. Once compression process is complete, a constant volume heat rejection process takes place (2–3), during which the piston reaches its TDC. In order to keep the volume constant, the expander moves back towards its BDC due to movement of the gas through the regenerator to the expander. During the gas flow through the regenerator, the matrix stores the heat from the gas, and thus pressure of the gas decreases due to cooling.

Further, the gas present in the regenerator and the expander is isothermally expanded (3–4), during which the expander piston reaches its BDC. This produces a cooling effect in the system. At steady-state condition, the temperature at the cold end of regenerator remains constant during this process, by virtue of the pressure decrease and the heat load on the system. Finally, the expander and compressor piston move back to their original positions, which causes the gas to move back to the compression space through the regenerator. The matrix rejects the heat taken by it back to the gas (4–1).

The ideal Stirling cycle is modified for its practical application, which demands the motion of the piston and displacer to be sinusoidal [8]. The refrigeration effect obtained in the Stirling cryocooler is due to the relative motion between the piston and the expander-displacer, wherein piston leads the displacer motion by a constant phase angle. For an efficient operation, this phase difference is maintained

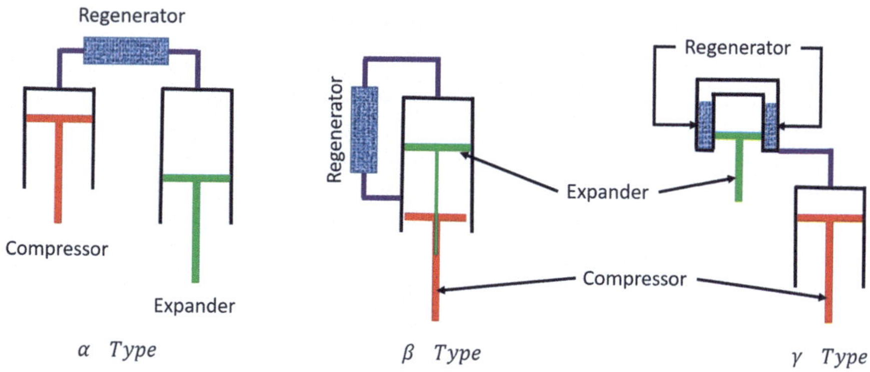

Fig. 1.5 Types of Stirling cryocoolers [5]

at an optimum value. Helium/hydrogen is the preferred working fluid for these cryocoolers, although use of mixtures has also been investigated [9].

The miniature Stirling cryocoolers that are generally used in space applications deliver a cooling effect of 1–5 W at around 80 K with a power input of around 20–50 W. The liquefier versions of Stirling cryocooler require around 1–2 kW of power input to get a liquefaction rate of approximately 1 litre per hour (lph) of liquid nitrogen. Depending upon the relative arrangements of compressor piston and expander-displacer, various configurations of Stirling cryocoolers are available. These are namely α-, β- and γ-type Stirling cryocoolers, as shown in Fig. 1.5.

Most miniature Stirling cryocoolers use oil-free linear compressors with flexure bearings. The liquefier versions (for nitrogen) of these coolers use oil flooded, crankshaft-driven compressors. However, linear compressor-driven liquefiers are also commercially available for laboratory purposes.

Since Stirling cryocoolers were the first of all type of cryocoolers to be commercialized, they are the most advanced and reliable of all the cryocoolers. They are high-efficiency machines, which operate at a frequency of 25–100 Hz. Miniature versions of such cryocoolers, which have a compact size and weight, are generally used for space applications. Since space applications demand very high reliability, these coolers are time tested and are a norm for many space-borne IR detectors. However, space applications prefer dry or gas lubrication, which demand very stringent tolerance requirements for these units. Therefore, the fabrication demands and cost of these units is high. In addition, Stirling cooler has vibrations, which can become a cause of concern for some specialized measurements. Sunpower® Inc., Thales Cryogenics and AIM INFRAROT-MODULE GmbH [10] are some of the commercial enterprises that offer miniature Stirling cryocoolers. The liquefier versions of Stirling cryocoolers use multi-cylinders and hence give a high cooling effect with an efficient operation. The capacity of Stirling cycle-based nitrogen liquefiers ranges between 5 and 100 lph. However, since these units are usually oil lubricated, they need periodic maintenance.

1.3.2 Gifford-McMahon (G-M) Cryocoolers

Gifford-McMahon cryocoolers are essentially derived from Stirling cryocoolers, as the cold head is similar for both the types. The major differentiating characteristic between a Stirling and a G-M cooler are the valved connection between the compressor and the expander units, rubbing low temperature seal at the cold end in case of a G-M cooler, and the frequency of operation. Figure 1.6 shows a schematic diagram of the G-M cryocooler.

The cold head of the G-M cooler, i.e. the regenerator, the expander-displacer and the seals are also shown alongside. The valve used in the G-M cryocooler is generally of a rotary type. It allows the high-pressure gas to enter the cold head during pressurization. Further, it releases the gas to low pressure side after producing a cooling effect at the cold end. The timing of the valves controls the displacer motion and hence allows it to operate at a lower frequency to that of the compressor piston. The valve of the G-M cooler typically operates at a frequency of 1–2 Hz. Therefore, the displacer of the G-M cooler requires good sealing that can withstand low temperatures. This seal is essential to prevent unwanted leakages of the working fluid between the displacer and the cylinder in the cold head.

In 1960, Gifford and McMahon demonstrated that by use of a rotary valve, variable pressure oscillation could be obtained by alternating between the high- and low-pressure sources [11]. In the 1980s, the G-M cryocooler-based cryopumps were commercialised for maintaining ultra clean vacuum for industrial applications. These coolers could reach temperatures of about 15–20 K. Further research was carried out in 1990s by use of rare earth regenerator matrix materials in the G-M cooler. The use of rare earth materials (having very high heat capacities at low temperatures) allowed the user to reach temperatures below 10 K [12]. These coolers were then employed in MRI and NMR systems for shield cooling to increase the

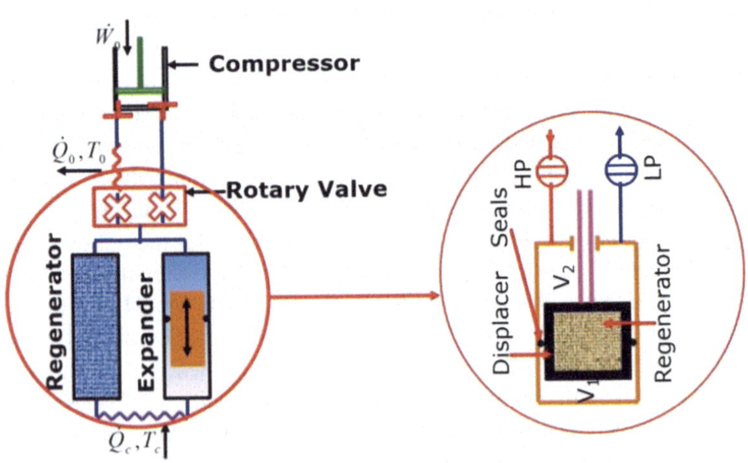

Fig. 1.6 Schematic diagram of G-M cryocooler [5]

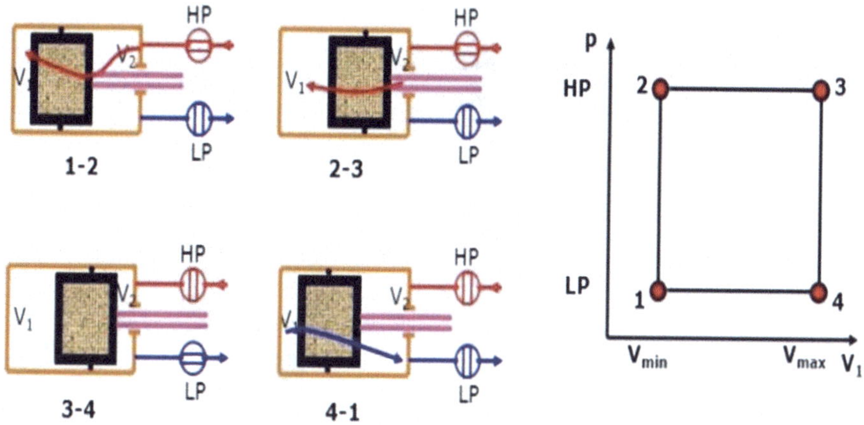

Fig. 1.7 Working principle of G-M cryocooler [5]

LHe hold time. The modern zero boil-off (ZBO) MRI systems use 4 K G-M-type cryocoolers.

As opposed to a Stirling cryocooler, the displacer of a G-M cryocooler does not move at the same frequency as that of the compressor piston. The working principle of the GM cryocooler can be better understood from Fig. 1.7.

It shows the movement of the displacer in the expansion chamber in relation to the opening and closing of the valves. Figure 1.7 also shows a graph of pressure vs. volume as per the displacer motion. Initially, both the high-pressure (HP) and low-pressure (LP) valves are closed, as shown in Fig. 1.7 (State 1).

In process 1–2, the HP valve is opened, which causes the pressurized gas to fill the spaces V_1 and V_2, which are the cold and the hot spaces of the regenerator, respectively. In process 2–3, the displacer moves back to displace the gas from V_2 to V_1 through the regenerator at a constant pressure, due to which V_1 increases in volume. This increase in volume is also noticed in the PV diagram in Fig. 1.7. In process 3–4, the HP valve is closed and the LP valve is opened, causing an expansion of the gas and hence reducing the pressure from HP to LP. This expansion produces a cooling effect in V_1. Later, in process 4–1, the displacer moves back to its original position, thereby reducing V_1. The low temperature gas moves back to the compressor after getting warmed in the regenerator. This cycle continues to produce lower and lower temperatures in subsequent cycles.

The rotary valve facilitates the production of any kind of pressure wave as per the requirement of system. The presence of the valve mechanism also allows the cold head to be placed at a distance from the compressor and connected by flexible lines for the high- and low-pressure gas.

Figure 1.8 shows the compressor and cold head of a commercial two-stage G-M cryocooler [13]. The figure shows a SHI®-make RDK-415D cold head with a water-cooled compressor. It gives a cooling effect of 1.5 W@4.2 K at the second stage with a power input of about 6–8 kW. The compressor of a G-M cooler may

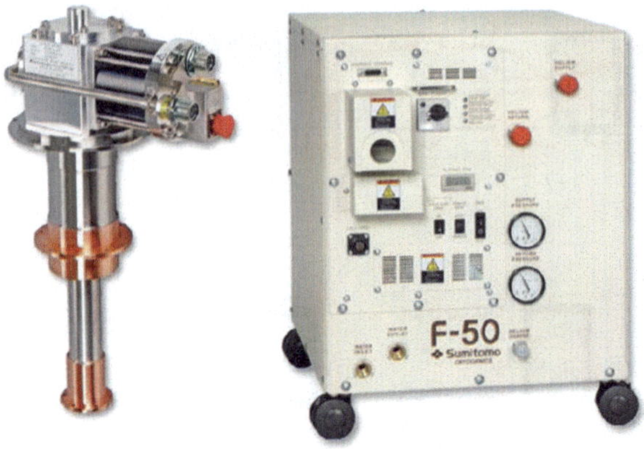

Fig. 1.8 RDK-415D cold head and F50H water-cooled compressor [13]

be air-cooled or water-cooled. The compressor and the cold head are connected by means of flexible hoses.

G-M cryocoolers are highly reliable and operate at low frequencies. On the other hand, they are less efficient, and the cold head has vibrations due to the displacer motion. The acoustic noise due to gas flow in flexible transfer lines from compressor to cold head is also an area of concern, especially in hospital environments for MRI applications. The rubbing seals in G-M coolers also pose maintenance problems.

G-M cryocoolers have a wide commercial value in applications such as cryopumping, cooling of thermal shields for superconducting magnet cryostats, and cooling of dry HTC SC magnets. Typically, a two-stage G-M cryocooler delivers a cooling effect of 0.1–1.5 W at 4.2 K with a power input of about 2–8 kW.

1.3.3 Pulse Tube Cryocooler (PTC)

The pulse tube cryocoolers (PTC) have a major advantage over the Stirling and G-M coolers due to absence of moving parts at the cold end. The moving displacer, which was prevalent in the Stirling and G-M cryocoolers, is replaced by an oscillating gas flow in a thin-walled tube. This tube is known as the pulse tube (PT), which contains the working fluid in the form of a gas. This is generally accompanied by a phase shift mechanism (PSM), which assists in producing a cooling effect in the PTC system. The pulse tube action may be perceived by assuming a gas displacer within the PT, which never leaves the PT.

A PTC may be of the G-M type or of the Stirling type. The valved PTCs are called as the G-M-type PTCs (GM-PTCs), whereas the PTCs without valves are the Stirling-type PTCs (S-PTCs). In many cases, the Stirling and G-M cryocoolers require a separate drive mechanism for the displacer, while the S-PTCs and G-PTCs do not need such mechanism. Generally, the miniature PTCs are of Stirling type, while the low-temperature PTCs, used for ground applications, are of the G-M type.

The idea of pulse tube cryocooler was introduced by Gifford and Longsworth in 1961 and was published in 1963. Gifford and Longsworth described the pulse tube cryocooler phenomenon as follows: ‘the pressurization and depressurisation of any closed volume from a point on its periphery sets up temperature gradients in the volume’ [14]. The development of PTC, in the true sense, was accelerated after about two decades by Mikulin et al. [15], when the phase shift arrangement (orifice) and reservoir were introduced. Further, Chen et al. [16] proposed a Double Inlet PSM for the G-PTC (DIPTC). After its potential for military surveillance was discovered, subsequent research and development was carried out by many researchers to improve various performance aspects of both types of PTCs [17–25]. A significant benchmark in PTC development was achieved by Matsubara et al. [17] in 1994, when they achieved temperatures below 4 K using a three-stage G-PTC. ER_3Ni was used as regenerator material in the last stage. Further, Heiden et al. [18] achieved a temperature of 2.07 K with a nitrogen precooled two-stage PTC. The use of G-PTCs for small-scale helium liquefiers was proposed by Wang et al. [19–21], which is now converted into a commercial liquefier by Cryomech®. In 2007, Yang and Thummus [22] proposed DI entry for S-PTC and obtained a temperature of 19.6 K with a two-stage PTC.

The complex phenomenon of compression and expansion of gas taking place in the pulse tube may be understood by taking into consideration the temperature distribution along its axis. Figure 1.9 shows a schematic of the Stirling-type pulse tube cryocooler and the temperature distribution along its length.

The components of a PT system are pressure wave generator (PWG), i.e. compressor, aftercooler, regenerator, cold end heat exchanger (CHX), pulse tube (PT), hot end heat exchanger (HHX) and phase shifting mechanism (PSM).

During pressurization, the high pressure gas flows across the regenerator and into the pulse tube. This gas compresses the existing gas in the PT, because of which temperature at hot end of PT increases in comparison with that of the bottom end.

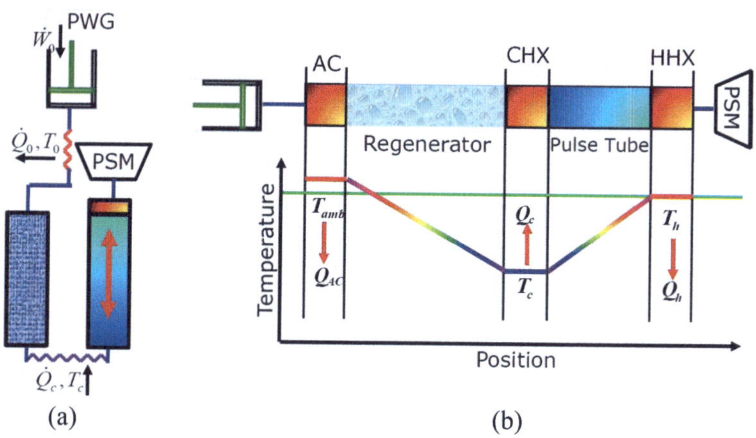

Fig. 1.9 Schematic diagram and temperature distribution along the length of a PTC

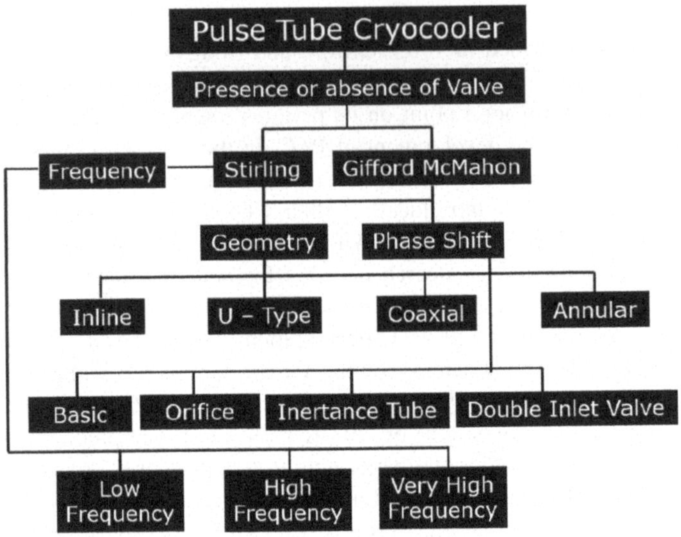

Fig. 1.10 Detailed classification of PTCs

During expansion, the temperature decreases and a temperature gradient is set up across the PT. The hot end of the PT is maintained at room temperature cooled by ambient or chilled water in the HHX. Hence, a refrigeration effect is obtained at the cold end of the PT. Apart from classification based on the presence and absence of valves, PTCs can also be classified based on different criteria. Figure 1.10 shows the detailed classification of PTCs along with the criteria used for classification.

1.3.3.1 Classification of PTCs Based on Geometry

Based on the geometrical arrangement of the pulse tube and the regenerator, the PTCs are classified into four configurations:

- Linear or inline
- U type
- Coaxial
- Annular

Figure 1.11 shows the different configurations of a PTC based on geometry.

In the inline arrangement (Fig. 1.11a), the regenerator and the pulse tube lie on the same axis. Thermodynamically, this may prove to be the most efficient configuration as it does not involve change in direction of the gas and has minimum void volume at the cold end. However, since the cold end is at the centre, it may not be easily accessible for some applications.

In U-type configuration (Fig. 1.11b), the regenerator and the pulse tube are placed parallel to each other and are connected by a 'U'-shaped tube. This has two effects:

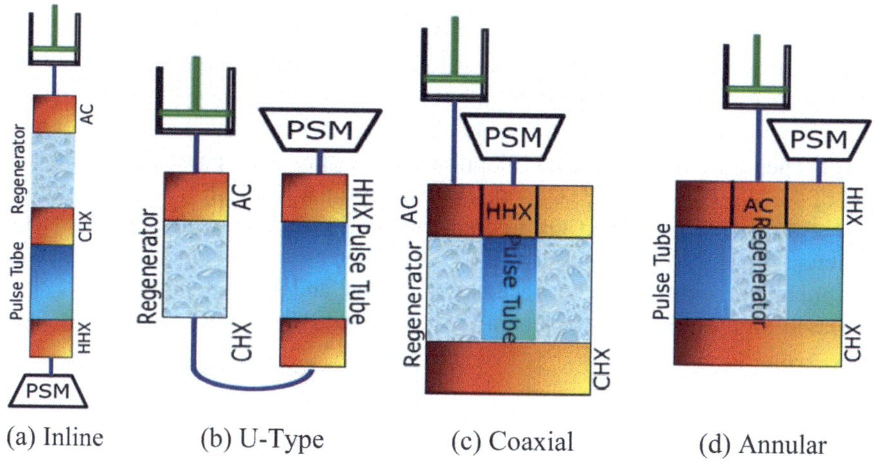

Fig. 1.11 Geometrical configurations of PTC. (**a**) Inline, (**b**) U type, (**c**) coaxial, (**d**) annular

void volume at the cold end of the pulse tube increases and pressure drop increases due to change in the direction of the gas. Both these effects reduce the pressure ratio of the system, thereby deteriorating its performance. However, the cold end is easily accessible for cooling purpose, and the PTC is more compact as compared to the inline configuration. Both the coaxial (Fig. 1.11c) and annular configurations (Fig. 1.11d) have either annularly placed regenerator or PT. Both these configurations are the most compact as compared to the other two configurations. However, the pressure drop is relatively more due to the sudden change in the direction at the CHX. In addition, there is a possibility of heat transfer between the PT and the regenerator which is undesirable.

1.3.3.2 Classification of PTCs Based on Phase Shifting Mechanism

In the Stirling or G-M cryocooler, the displacer motion is 90° out of phase with the compressor motion. Equation (1.1) gives the net cooling effect in a PTC in terms of mass flow rate at the cold end and the phase difference between mass flow rate and pressure pulse [24].

$$Q = \frac{1}{2} R T_c \mid m_c \mid \cos\theta \qquad (1.1)$$

where

Q = net refrigeration effect (W)
R = gas constant (J/kg K)
T_c = temperature at the cold end (K)
m_c = average mass flow rate at the cold end (g/s)

θ = phase angle between mass flow rate and the pressure pulse

From this equation, it may be noticed that a larger mass flow rate at the cold end of the PT and a smaller phase difference result in a greater net refrigeration effect. Theoretically, the phase angle should be zero to obtain the maximum cooling effect. Therefore, there is a need to optimize or minimise the phase shift, which is one of the major aspect of PTC modelling. An optimal phase shift ensures optimal size of the compressor and COP to obtain the required cooling effect of the PTC. Different phase shift mechanisms (PSMs) have been proposed in literature to obtain an optimal phase difference between mass flow rate and pressure pulse.

Based on these PSMs, Matsubara [25] classified PTCs into the following four types:

- Basic
- Orifice
- Inertance tube
- Double-inlet valve

Figure 1.12 shows the schematic diagram of all these PTCs.

For a BPTC, there is no phase shift mechanism. As a result, very low cooling effect is obtained at relatively higher temperatures. This is the reason why the BPTCs are redundant in the current scenario. The phase shifting devices commonly used in a PTC nowadays are orifice, double-inlet (DI) valve, and inertance tube (IT), or a combination of these types. A combination of two or more phase shifting devices is also used in practice.

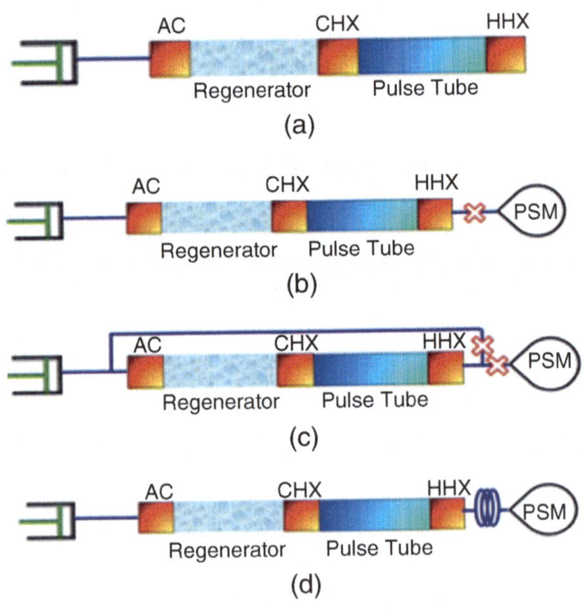

Fig. 1.12 Schematic of various phase shifting mechanisms of PTC. (**a**) Basic PTC, (**b**) OPTC, (**c**) DIPTC, (**d**) ITPS

1.3.3.3 Classification of PTCs Based on Operating Frequency

Based on operating frequencies, the Stirling PTCs are classified into three categories:

- Low frequency (<30 Hz)
- High frequency (30–80 Hz)
- Very high frequency (>80 Hz)

High frequency and very high frequency PTCs are generally used for space and military surveillance.

1.3.3.4 Advantages, Limitations and Applications of PTC

Due to the absence of a moving displacer in PTCs, there is no rubbing contact and no requirement of rubbing seals. Thus, PTCs have high reliability, longevity and better vibration characteristics. However, the energy dissipation in phase shifting mechanism reduces the COP of the cycle. In addition, as more gas is required to pass through the pulse tube and reservoir, the viscous losses increase. Another major limitation of PTCs is that its performance is orientation dependent.

PTCs are used in many applications, where the inherent simplicity of the PTC finds ready acceptance. Major applications of PTCs include recondensing of LHe and LN_2 in magnetic resonance imaging (MRI) and nuclear magnetic resonance (NMR) machines, storage of biological cells and specimens, etc. High frequency or S-PTCs are developed for space applications, which include cooling of infrared sensors for satellite-based surveillance, whereas ground applications mostly use GM-PTCs to reach lower temperatures.

1.4 Regenerators

Being the primary heat exchange medium for regenerative cryocoolers, the regenerative heat exchanger (HX) or regenerator is the most vital component for such cryocoolers. The effectiveness of the regenerator is a highly influencing parameter of the net cooling effect obtained from any cryocooler. Hence, the regenerator design optimization is a very important step for the design of such cryocoolers. This optimization is carried out based on the desired pressure drop and the heat transfer characteristics. The input parameters for the design of a regenerator are the working pressures and temperatures and the desired cooling effect.

For a high heat transfer rate at low temperatures, a finely divided mesh is required to increase the heat transfer area. In addition, the material must possess high heat capacities, especially at lower temperatures. This is because, higher heat capacity of the regenerator leads to higher effectiveness of the regenerator [12]. The porosity of the mesh and regenerator dimensions, i.e. length and diameter are optimized to obtain high effectiveness of the regenerator.

Based on the desired temperature range at which cooling effect is desired, the regenerator materials may be divided into three temperature ranges, 50, 10 and 4 K materials. Figure 1.13 shows the variation of volumetric heat capacity of various regenerator materials with temperature [26].

For single-stage cryocoolers operating in the range of 30–100 K, materials like stainless steel (SS) and phosphor bronze are used. However, these materials have lower heat capacity at temperatures below 20 K. Therefore, they are not preferable as low temperature regenerator materials. For 10 K cryocoolers, two or more stages are needed to obtain cooling effect at low temperatures. In such cryocoolers, the first-stage regenerator generally consists of SS meshes, while the second-stage regenerator is made of lead (Pb) balls. However, its heat capacity drops drastically at temperatures below 10 K [12].

For 4 K cryocoolers, rare earth magnetic regenerator materials, which have high specific heat capacities at temperatures below 10 K, are used. These include Er_3Ni, $HoCu_2$ and GOS and their alloys.

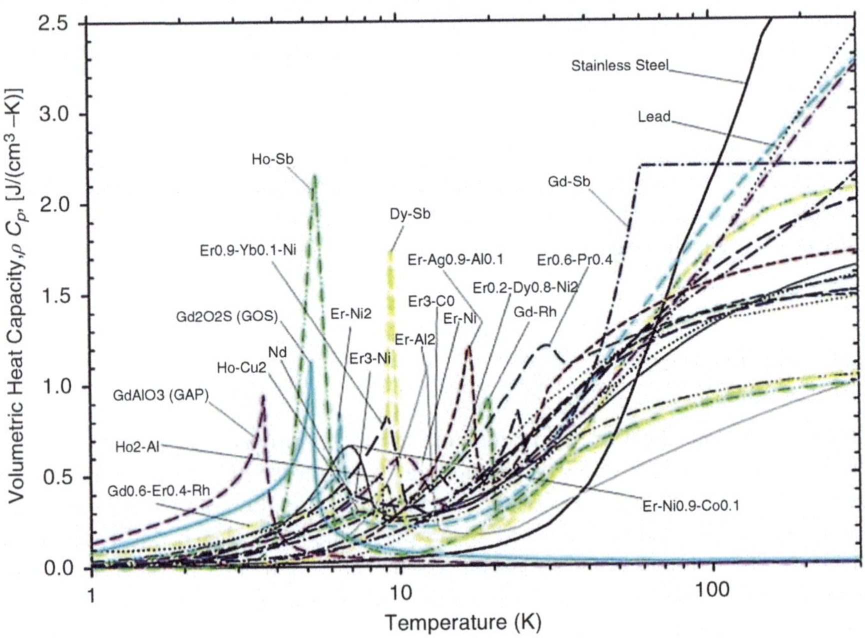

Fig. 1.13 Variation of volumetric heat capacity with temperature for regenerator materials [26]

1.5 Compressor for Regenerative Cryocoolers

Cryocoolers generally consist of two distinct elements: the compressor and the cold head. In closed cycle regenerative cryocoolers, the compressor delivers a pressure pulse used to drive the cold head. These compressors may be broadly classified as dry compressors or oil-lubricated compressors. These may be air or water cooled, depending on the audible noise and application requirements. Dry compressors are generally used to drive for Stirling and S-PTCs that are used in space and military applications.

These cryocoolers are of high-frequency type and use linear compressors. Since they move at such high frequencies, the gas forms a thin layer which acts as a lubricant and minimises the rubbing action of the displacer. A high-frequency linear compressor was developed at the University of Oxford by Davey et al. in 1990, and several studies were carried out to reduce the vibrations in the developed system [27–33]. Figure 1.14 shows a schematic diagram of a dual opposed piston moving-magnet compressor [34]. A CHART® Inc. linear compressor of the same configuration is shown alongside [35].

However, they have lower power rating as compared to oil-lubricated compressors. The liquefier versions of the Stirling cryocoolers use oil-lubricated compressors. These compressors are of the crankshaft-driven type and supply high input power to the cryocooler to deliver better cooling effect for high liquefaction rates.

G-M-type cryocoolers typically use oil-lubricated compressors, which may be air or water-cooled.

Figure 1.15 shows air and water-cooled helium compressors that are used to drive G-M or G-PTCs [13]. G-M coolers have low COPs due to the presence of rotary valve. It requires high input powers of around 8 kW. Since these are oil-lubricated,

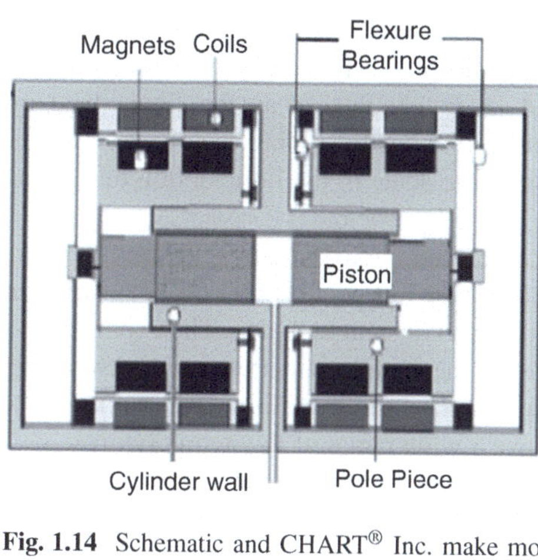

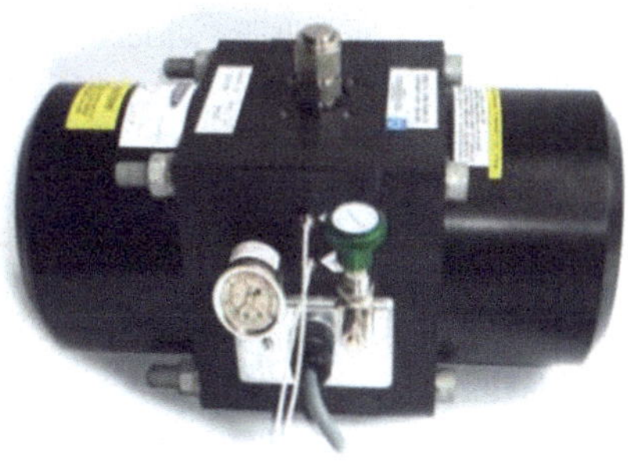

Fig. 1.14 Schematic and CHART® Inc. make moving magnet dual-opposed piston linear compressor [34, 35]

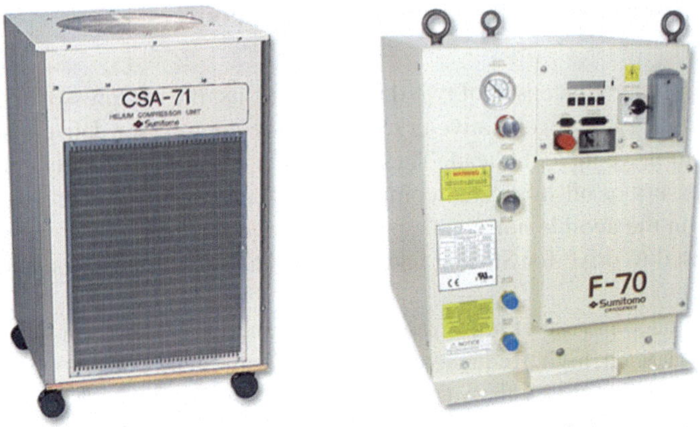

(a) (b)

Fig. 1.15 Sumitomo® air and water-cooled helium compressors [13]

they require oil traps and filters and adsorbers to prevent the oil from contaminating the cold parts of the cryocooler. The working fluid for these compressors is helium (He).

A lot of research is being conducted all around the world for high-efficiency linear compressors.

In addition, alternative static drives such as thermoacoustic and sorption compressors are also being investigated.

1.6 Joule–Thomson (J-T) Cryocoolers

Application of Joule–Thomson (J-T) effect in gas liquefaction cycles was discovered by Carl von Linde and William Hampson [36]. J-T cryocooler is the simplest type of cooler which is widely applied as cryocooler or as the final stage of most liquefaction cycles. The cryocooler can be easily miniaturized, while the J-T expansion process is also used on a very large scale in the liquefaction of natural gas or even for other gases [37].

J-T cryocoolers work in either open or closed cycle, to attain temperatures in the range of 75–120 K. In J-T cycles, the expansion occurs with no heat input or production of work; thus, the process occurs at a constant enthalpy. The expansion of gas results in lowering of temperature only if the initial temperature of the working fluid is below its characteristic temperature, called the inversion temperature [38]. Different working fluids are used to achieve a wide range of low temperatures.

Figure 1.16 shows the principle of operation of a typical closed cycle J-T cooler.

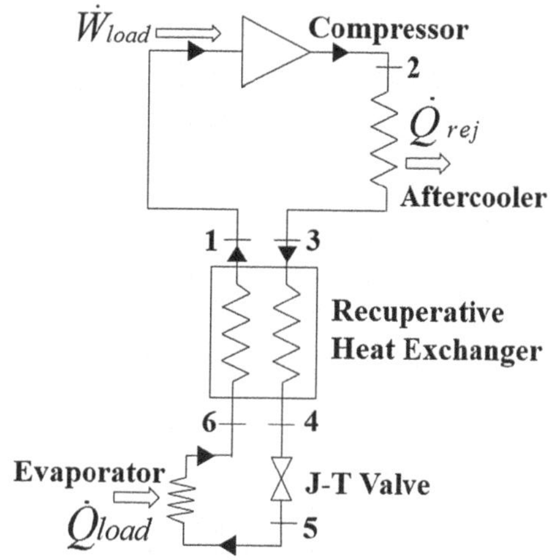

Fig. 1.16 Schematic diagram of a Joule–Thomson Cryocooler [37]

The major components of closed cycle J-T cooler are a compressor, an aftercooler, a counterflow heat exchanger, an expansion/throttle device (J-T valve) and a cold end/cold finger. A compressor is used to compress the working fluid (typically nitrogen or argon) to a high pressure of 20 MPa or more on the high pressure side (1–2) to achieve reasonable cooling. The high pressure stream is cooled to an ambient temperature by the aftercooler (2–3). This high pressure stream then passes through a counterflow heat exchanger, where it is precooled prior to the J-T expansion by the returning low pressure, low temperature gas stream from the evaporator (3–4). The high pressure fluid then undergoes an isenthalpic expansion in a J-T valve, cools down and changes partially its phase to a liquid (4–5). The expanded liquid absorbs the heat load at constant temperature in the evaporator (5–6) and then returns to the compressor through the heat exchanger (6–1) to complete the cycle.

The use of mixture as a working fluid within a J-T cycle increases the J-T effect since the vapour dome associated with a mixture of gases tends to extend over a larger temperature range as compared to that for a pure gas. Although the use of mixed refrigerants (MR) was proposed by Podbiclnak in 1936 for liquefaction of natural gas [39], its application for small-sized J-T coolers was developed in the 1970s and 1980s in Soviet Union projects. Often known as Kleemenko coolers, these mixed refrigerant J-T cryocoolers (MRJT) are found to improve the thermodynamic efficiency of single-fluid J-T coolers by an order of magnitude. They have a lifetime of about 8–10 years [40].

Figure 1.17 shows a temperature (*T*)–enthalpy (*h*) diagram of a J-T cycle for pure nitrogen and a nitrogen hydrocarbon mixture for typical values of working pressures.

For the sake of simplicity and comparison, the compression process is not shown in the *T*–*h* diagram. For single-fluid J-T cryocoolers, the required pressure ratio in order to obtain cooling effect at cryogenic temperatures is about 100–200, whereas the same for MRJT coolers is about 10–20. In addition, the high- and low-pressure

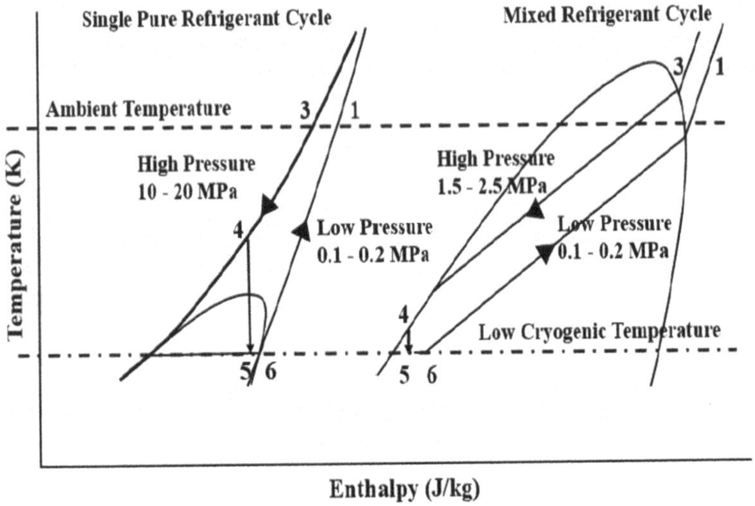

Fig. 1.17 Comparison of single-refrigerant and mixed-refrigerant J-T cycle [41]

streams passing through the heat exchanger for a single-fluid J-T cycle are in gaseous phase, whereas these are in the two-phase region of a MRJT cycle. This results in higher heat transfer coefficients and better heat exchange effectiveness in MRJT cycles.

Furthermore, a highly efficient cycle yielding significant refrigeration is possible with high pressures of about 2.5 MPa for MRJT cycles, which are easily achievable by using conventional domestic refrigerator compressors, thereby reducing costs. All these factors have influenced the development of MRJT cryocoolers.

J-T cryocoolers are simple in construction, have a compact structure and can have rapid cool down characteristics. The main advantage of J-T cryocoolers is that there are no moving parts at the cold end, which allows them to be miniaturized. Hence, the reliability of the system is mainly decided by the reliability of the compressor. In addition, clogging of expansion devices is also possible in such systems if not purged properly. The thermodynamic efficiency of these systems is low as compared to isentropic expander systems. The main factor responsible for the low efficiencies is the process irreversibility in the compression, in the heat transfer process taking place in the heat exchanger and in the throttling process.

In addition, the cold end can be separated from the compressor by a large distance, greatly reducing the electromagnetic interference and vibration originating from the compressor. Single-fluid J-T coolers are used in various applications to attain cooling to near LN_2 temperatures of IR detectors for missile guidance, cooling of gamma ray detectors, cryosurgical probes, precooling of helium in Collins cycle-based helium liquefiers, while MRJT coolers are used for liquefaction of natural gas, storage of biological cells and specimens, etc.

1.7 Design and Analysis of Cryocoolers

Due to the various losses involved in the system, especially at cryogenic temperatures, every cryocooler must have an optimized design in order to deliver the required cooling effect. Since cryocoolers are specialized devices, a universal design procedure for each type of cryocooler, which optimizes all the design considerations and maximizes the COP of the cooler, is difficult to establish. Hence, several methods exist in literature for the design of different types of cryocoolers, each of which focuses on optimizing different aspects of cryocooler performance. A basic thermodynamic analysis of all types of cryocoolers has been given by deWaele [42].

1.7.1 Stirling Cryocooler Analysis

The first-order analysis methods for Stirling cryocoolers include Schmidt analysis and Walker's optimization charts [1]. Although these are parametric studies based on several idealization assumptions, these methods give a significant insight towards the performance of the Stirling cryocooler. Schmidt analysis is based on a realistic cycle and gives an initial guess of the cryocooler dimensions based on the input parameters. It is used widely as an approximation for the refrigeration effect for Stirling cryocooler. The performance of the cryocooler can also be studied against the variation of each parameter individually by using Schmidt analysis method. In order to find optimum dimensions of a cryocooler, a compromise must be sought between all design parameters. Based on the parametric study from Schmidt's analysis, a combined effect of these parameters on the performance of system as a whole, was first reported by Graham Walker in the year 1962 in the form of design charts. These charts were produced for both refrigerators and engines and are now called as Walker's optimization charts. However, all losses and inefficiencies involved in the actual operation are not taken into consideration to calculate the net refrigeration effect in these methods. To overcome these limitations, various numerical models have been published in literature for Stirling cryocoolers [43–61]. Each of these models focuses on various aspects of the cryocooler performance.

The aim for all of these analysis methods is to obtain an optimal combination of design parameters in order to get maximum refrigeration effect at the minimal required power input [50]. Hence, most of them are related either to regenerator parameter evaluation and optimization or on compressor design or pressure wave generator parameters [45–53]. Over the past 20 years, various types of Stirling cryocoolers have been developed, which are driven by crankshaft, free piston free displacer or free displacer compressors or pressure wave generators [62]. The mechanism and characteristics of these cryocoolers have thus been studied in detail by various researchers over these years. Atrey et al. [43] used an adiabatic model for compressor and expansion volumes for analysis of several Stirling cryocoolers, after which Bauwens [57] divided the cryocooler into various spaces, which were either adiabatic or isothermal. Several mathematical models and performance studies have been conducted over the years on various aspects, which can be broadly classified into regenerator parameters evaluation and optimization, cryocooler performance studies in various operating conditions or for particular applications, or compressor parameter optimization and evaluation [49–64]. Recent methods also involve CFD analysis of the Stirling cryocooler and their validation with experiments. Currently, there is a lot of research on the analysis of miniature high-frequency Stirling cryocoolers [55, 63] as well as compressors for Stirling cryocoolers [64].

1.7.2 G-M Cryocooler Analysis

Few analytical methods have been developed to study the performance of GM cryocoolers [65–68]. However, these methods do not take into consideration the actual pressure and mass flow rate changes in the cycle. A heat balance analysis of the GM cooler [69–75] for single- and two-stage cooler does not take into account the various losses in the system. Some dynamic models are also available in literature that represent the physical trends with reasonable accuracy [72, 73]. A few comprehensive numerical models for GM cooler are available in literature for a complete GM cooler analysis [74, 75]. As the pressure pulse remains more or less constant in the system, this is generally given as input to the simulation. As far as GM coolers are considered, most of the research studies nowadays focus on the performance of different regenerator materials, which has been dealt with in the previous section.

1.7.3 Pulse Tube Cryocooler Analysis

Various basic analysis methods are available in literature for different types of PTCs [76–78]. The phasor analysis given by Radebaugh [79] is essential to develop a good understanding of the working of different types of PTCs. Based on this methodology, the different types of PTCs have been analysed by this method [80, 81]. Phasor analysis is a vectorial representation of mass flow rates, pressures and temperatures at different locations of the PTC as a function of time. According to phasor analysis, one can obtain the net refrigeration effect of the PTC in terms of mass flow rate at the cold end of the PT, phase difference between the mass flow rate and the pressure pulse. The net refrigeration effect is given by Eq. (1.2) or Eq. (1.3) as

$$Q = \frac{1}{2} R T_{\mathrm{c}} \frac{P_{\mathrm{i}}}{P_{\mathrm{o}}} \mid \dot{m}_{\mathrm{c}} \mid \cos\theta \tag{1.2}$$

$$Q = \frac{C_p}{2} \frac{P_\mathrm{i}}{P_\mathrm{o}} T_\mathrm{h} \mid \dot{m}_\mathrm{c} \mid \cos\theta \frac{T_\mathrm{c}}{T_\mathrm{h}} \tag{1.3}$$

where

Q = net refrigeration effect, W
R = gas constant, J/kg K
T_c and T_h = cold end and hot end temperatures, K
P_i and P_o = inlet and outlet pressures, bar
$\dot{m}_\mathrm{c}$ = mass flow rate at the cold end of the PT, kg/s
θ = phase difference or phase angle
R = gas constant, J/kg K
C_p = specific heat capacity at constant pressure, J/kg K
γ = adiabatic constant for working fluid

This vectorial representation of the above equation is shown in Fig. 1.18.

Based on the phasor analysis, the phasor diagrams of the BPTC, OPTC, DIPTC and ITPTC (representative) are as shown in Fig. 1.19.

A lumped electric model proposed by Gardner and Swift [23] for PTC is also another first-order analysis method. The analogies between the process parameters and electrical quantities defined by the model are given below:

- Oscillating pressure (p) analogous to voltage (V)

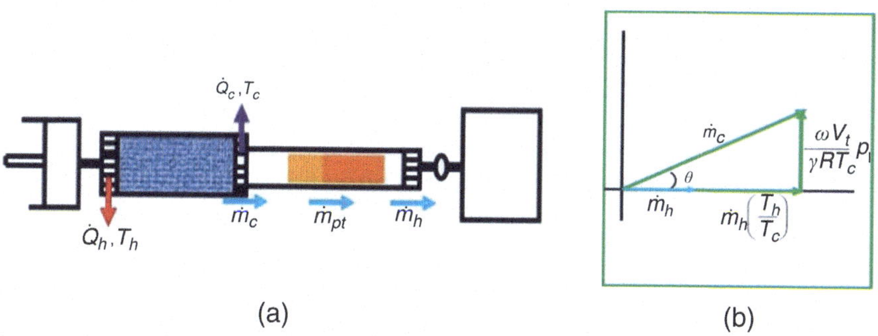

Fig. 1.18 Schematic and phasor diagram of a PTC. (a) Schematic of PTC. (b) Phasor diagram of PTC

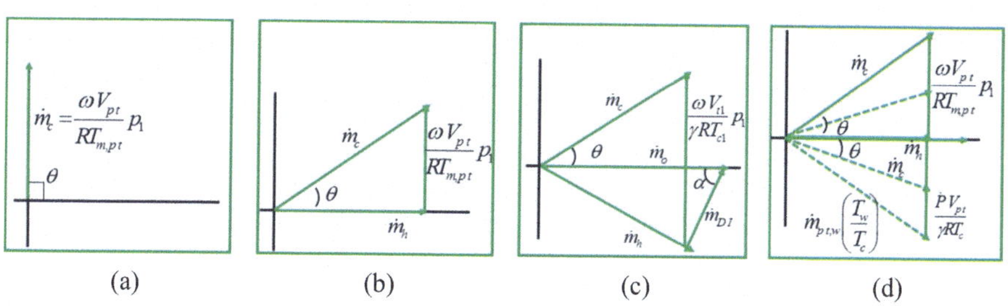

Fig. 1.19 Phasor diagrams of (a) BPTC, (b) OPTC, (c) DIPTC and (d) ITPTC

- Mass flow rate (m_c) analogous to current (I)
- Reservoir volume analogous to capacitance (C)
- Orifice resistance analogous to resistance (R)
- Inertance analogous to inductance (L)

This electrical analogy is useful to analyse the working of a PTC as an electrical model and the function of the inertance tube.

The second-order methods for PTCs are simple computational numerical models [82–86]. Out of these, the isothermal model given by Zhu and Chen [82] is the most commonly used model to analyse PTCs. The isothermal model is a very simple and useful methodology used to analyse Stirling cryocoolers. The second-order isothermal model given by Atrey et al. [83] combines the cyclic analysis and the isothermal model of Zhu and Chen to design and analyse the performance of two-stage Stirling-type PTCs.

The nodal analysis given by Zhu and Matsubara [87] uses continuity, momentum and energy equations of gas for the analysis by appropriate changes in boundary conditions depending on the type of PTC. Other third-order numerical models have been given in literature by Wang et al. [88, 89], Ju et al. [90], Tau et al. [91] and many other researchers.

Several CFD analyses are also available for PTCs [92–103]. In addition, Sage software [103] is commercially available and widely used to model and optimize Stirling coolers and PTCs [104–107]. Sage is based on a simulation model that facilitates user-defined specifications such as component dimensions and optimizes all parameters to give the best prediction of performance of the cryocooler within a graphical user interface.

1.7.4 Joule–Thomson Cryocooler Analysis

For single-fluid J-T coolers, the analysis is mainly focussed on the heat exchanger performance, whereas for the MRJT coolers, in addition to heat exchanger performance, the mixture optimization and correlation development for local heat transfer coefficients are also significant aspects of research.

The theoretical work on the single-fluid J-T cryocooler is scarce due to complex geometry of the Hampson-type heat exchanger, thermal losses and variable physical properties of compressible fluid. Maytal [108] analysed the performance of an ideal flow-regulated Hampson type J-T cooler. Xue et al. [109] and Ng et al. [110] reported experimental and numerical study of the J-T cryocooler for steady-state characteristics with argon (Ar) as a working fluid. Hong et al. [111] used an effectiveness NTU approach to predict the performance of the heat exchanger for pressures up to 500 bar with argon and nitrogen as working fluids.

Recently, Ardhapurkar and Atrey [112] presented the steady-state performance analysis of a miniature single-fluid J-T cryocooler. They determined the design and operating parameters for the optimum performance of the cryocooler. They also studied the effect of various parameters such as supply pressure, fin density and mass flow rate on the performance of the cryocooler.

Chou et al. [113] conducted studies on transient behaviour of J-T cryocooler for a fixed orifice that does not contain a flow-regulating mechanism. Chien et al. [114] presented a numerical simulation of the transient characteristics of cool-down process for a self-regulating J-T cryocooler. More recently, Damle and Atrey [115] computed the transient performance of a miniature J-T cryocooler. They considered the distributed J-T effect in the numerical simulation of the recuperative heat exchanger. In this study, it is demonstrated that the distributed J-T effect is pronounced when the pressure drop on the high-pressure side of heat exchanger is substantial. As a result, there is a significant impact on the temperature profile, cold end temperature and the cool-down time of the cryocooler.

1.7.4.1 Performance Analysis of MRJ-T Cryocooler

The performance of a MRJ-T cryocooler depends on the gas mixture used, the design of the heat exchanger and the compressor. The recuperative process in the J-T cryocooler operating with multicomponent mixture is one of the most important processes. Most thermal losses occur in the recuperative heat transfer process, which influence the performance of the whole system. There are a few studies related to multicomponent mixtures boiling at cryogenic temperatures. As a result, design of these heat exchangers is a challenging task.

Non-linear variation of enthalpy with temperature in the two-phase zone causes variation in temperature difference in the recuperative heat exchanger. The study of temperature and pressure profiles in the heat exchanger and their dependence on the mixture composition is crucial for the optimization of the mixture. Experimental data and empirical correlations for predicting heat transfer coefficients of evaporating and condensing streams of multicomponent mixtures are not available in the literature. Therefore, the thermodynamic performance of the heat exchanger is analysed using Q-T diagram, which is a plot of temperature versus the heat transferred. Gong et al. [116] and Ardhapurkar et al. [117] studied the effect of mixture composition and their properties on the temperature profiles in the heat exchanger using Q-T diagrams. Ardhapurkar et al. [117] presented the flowchart for the computational analysis developed to obtain temperature distributions with respect to heat transfer. The input parameters to the analysis are the mixture composition and properties, operating pressures (HP and LP), inlet temperature of the hot and the cold fluid and the heat exchanger pinch point temperature difference.

1.7.4.2 Mixture Optimization

Gases such as helium, neon, nitrogen, argon, methane, ethane, propane, butane, CF4 and HCFCs can be constituents of the mixture refrigerants. Also, the possible range of percentage of the each component in the mixture is wide, as reported in different mixtures. It mainly depends on the required low temperatures and cooling capacity. However, there is no unique mixture composition for a given temperature/capacity application. Therefore, selection of components and their composition is very significant, and it decides the performance of a system. There are no specific rules for selecting the mixture composition. Venkatarathnam [118] and Walimbe et al. [119] suggested the guidelines to select mixture components so as to produce different range of refrigeration temperatures. Many researchers [37, 41, 120] optimized the mixture composition by maximizing the minimum isothermal enthalpy difference between the high- and the low-pressure streams over the complete operating temperature range of the heat exchanger.

Mostly, the mixture compositions used in the literature are patented for specific applications. Keppler et al. [37] used a genetic algorithm to optimize the mixture for a specified set of components, operating conditions and load temperatures suitable to a single-stage J-T cryosurgical system. Little [120] patented the optimization method for liquid refrigerant supply (LRS) systems in order to maximize exergy efficiency and heat transfer capacity of a heat exchanger. Alexeev et al. [41] presented the method to optimize the mixture composition and to determine the high and the low pressures for the required cooling temperature. Gong et al. [121] developed a complex optimization algorithm to optimize the mixture composition for a Linde–Hampson refrigerator. Alexeev and Quack [122] optimized the mixture composition and patented it for a pre-cooled refrigerator operating in a temperature range of 90–110 K. The specific refrigeration effect is maximized in their method similar to that described by Radebaugh [123].

1.7.4.3 Correlation for Local Heat Transfer Coefficients

Many researchers [124–127] analysed the performance of heat exchangers in terms of overall heat transfer coefficients, which does not allow investigation of the local heat transfer coefficients.

Gong et al. [128] presented the pool boiling data for methane and its natural gas mixtures. Nellis et al. [129] obtained experimental data of heat transfer coefficients for mixed refrigerants used in the cryocooler at various operating conditions. This is probably the only reported study on the flow boiling of nitrogen–hydrocarbon mixtures at cryogenic temperatures. On the other hand, there is no experimental study found on local values of condensation heat transfer coefficients of cryogenic mixed refrigerants. On the basis of the data reported by Nellis et al. [129], Little [130] derived the correlation for two-phase heat transfer coefficient which is closely related to the Silver [131] and Bell-Ghaly equation [132]. Recently, Ardhapurkar et al. [133] assessed the existing flow boiling correlations for mixtures of nitrogen–hydrocarbons using experimental data reported by Nellis et al. [129]. It is found that the existing pure component correlations are not suitable for the mixed refrigerants. They modified Gungor and Winterton [134] and Granryd correlation [135] to apply them for multicomponent mixtures at cryogenic temperatures. Ardhapurkar et al. [136] also studied the applicability of the existing pure component condensation correlations to mixed refrigerants at cryogenic temperatures. The calculated condensation heat transfer coefficients are corrected using the Silver-Bell-Ghaly method to take into account the effect of mixture and recommended Cavallini and Zecchin correlation [137] for the calculation of condensation heat transfer coefficients of the mixed refrigerants at cryogenic temperatures.

1.8 Worldwide Scenario

Continuous research on closed cycle cryocooler has improved its efficiency over the years. This has enabled its usage in a multitude of applications, leading to an increase in its production quantity, thereby reducing its costs. Figure 1.20 gives most of the commercial applications of cryocoolers [138].

The applications of cryocoolers may be broadly classified under 'big' and 'small' cryogenics, depending on the physical size of the application. The use of cryocoolers in large space simulation chambers and accelerators has given a major boost to the cryocooler production rates in the past three decades. These applications can be clubbed under 'big' cryogenics. Figure 1.21 shows a large cryopump used to maintain vacuum in a large space simulation chamber [139].

'Small cryogenic' includes applications such as MRI/NMR systems, low capacity liquefiers, IR detectors, cryopumps in semiconductor industry and other high-vacuum environments, HTS (high-temperature superconductor) and LTS (low-temperature superconductor) superconducting devices, NMR probes, cryogen-free magnets and dilution refrigerators.

Apart from the commercialized cryocoolers, extensive research is being carried out to improve the COP and mean time to failure (MTTF) of the cryocoolers to have better reliability. In addition, research efforts are also concentrated to eliminate its

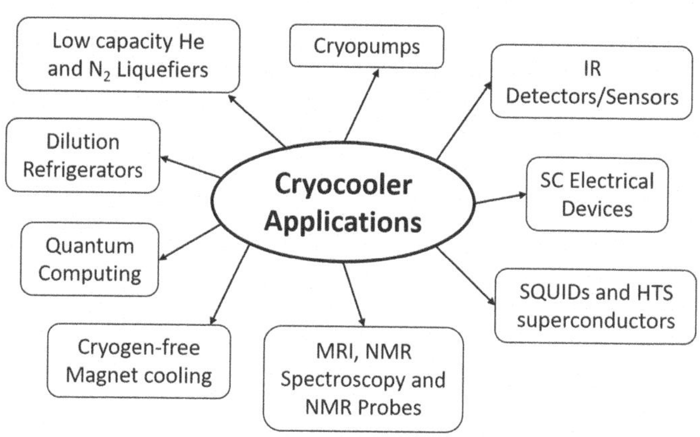

Fig. 1.20 Commercial applications of cryocoolers

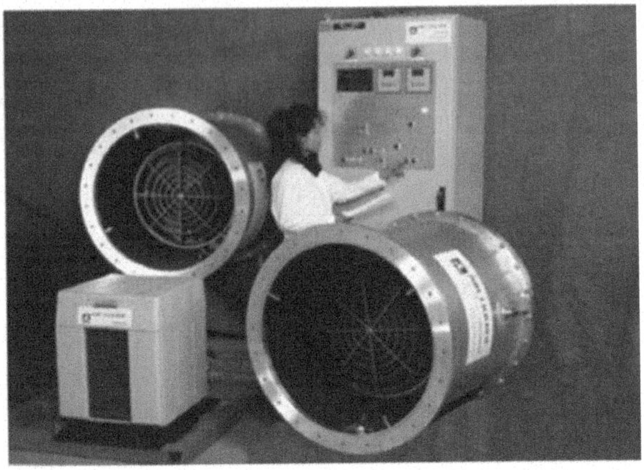

Fig. 1.21 Cryopump for a space simulation chamber [139]

inherent disadvantages such as audible noise, electromagnetic interference (EMI) and vibrations. This section highlights the commercial cryocoolers used in industry and also discusses the current research activities that are being conducted worldwide. Conductively cooled SC magnets and recondensing systems for accelerators are still in the nascent research phase due to the enormity and scale of operation. Therefore, the available accelerators still follow the conventional way of cooling with a liquid helium bath. Since small cryogenic applications are more closely linked to cryocooler developments, only these aspects are discussed below in detail.

1.8.1 Cryopumps

Since 1980s, the largest commercial demand for cryocoolers is for cryopumps [2]. Cryopump is basically a vacuum pump that captures the gas by cooling a surface to cryogenic temperatures. In order to attain various levels of vacuum, several pumping mechanisms may be employed. However, cryopumping is by far the most effective method to achieve a clean and constant high vacuum. 10 K G-M cryocoolers or cryogen cooled baths are generally used for most cryopumping applications. A cryopump works on the principles of cryotrapping and cryocondensation to provide high-vacuum environments [139]. The cold surfaces must be cooled to a temperature, so as to keep the corresponding saturation pressure equal to or below the desired vacuum pressure in the chamber.

Figure 1.22 shows the saturation pressure variation with temperature for some common gases [139].

It may be seen from this figure that, at a temperature of about 20 K, the saturation pressure of N_2, CO, H_2O, Ar, O_2, and CH_4 is below 10–10 Pa. Hence, these gases get freeze on the cold surface when it is maintained at 20 K. However, He, Ne and H_2 require temperatures of the order of 1 K and below to be cryo-condensed on the cold surface, which is quite difficult to achieve. In order to trap these gases, charcoal adsorbers are placed on the second-stage cold end of the cryocooler.

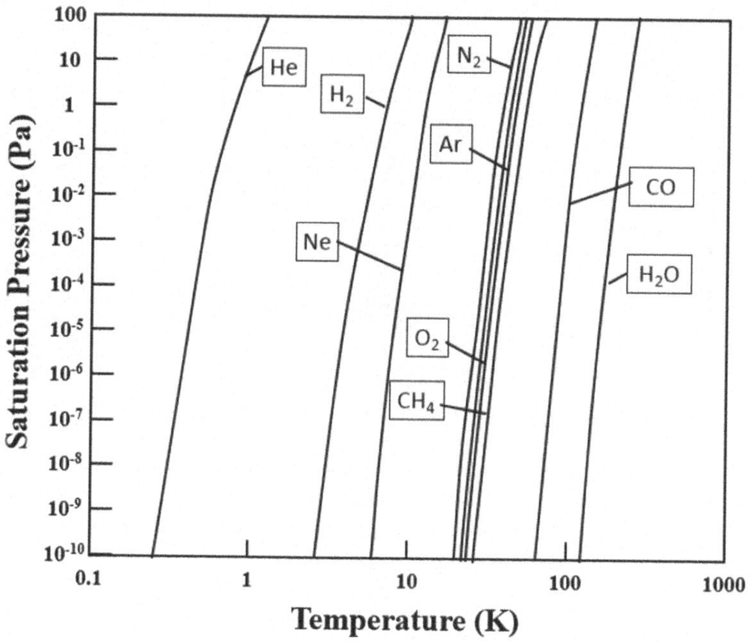

Fig. 1.22 Saturation pressure vs. temperature for common gases [139]

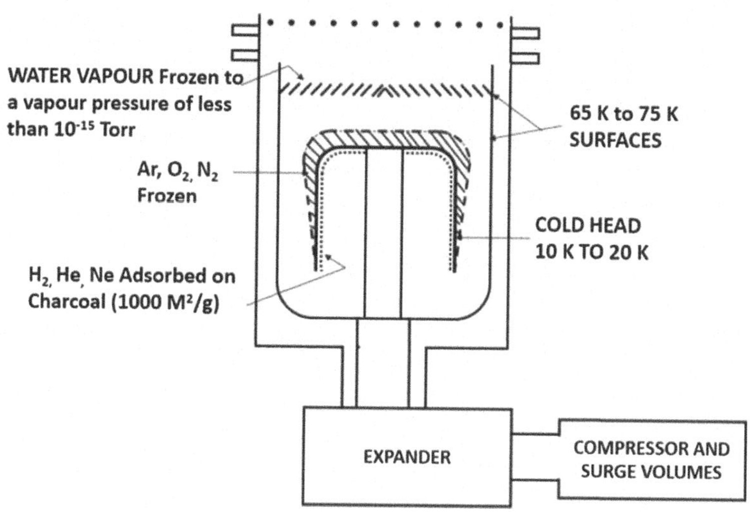

Fig. 1.23 Schematic and working principle of a cryocooler-based cryopump [139]

Figure 1.23 shows the working principle of a cryocooler-based cryopump. It shows which gases are trapped or condensed with the corresponding location in the cryopump.

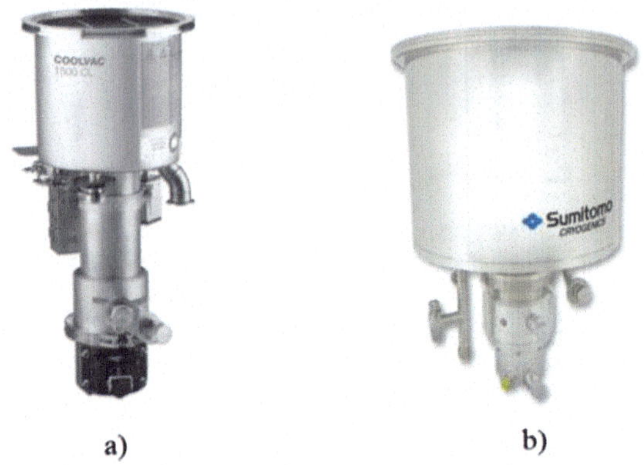

Fig. 1.24 Commercial cryopumps [13, 140]. (**a**) COOLVAC 1.500 CL automatic control cryopump by Leybold®. (**b**) Marathon® CP-16 Cryopump by Sumimoto®

Figure 1.24 shows SHI® and Leybold® cryopumps (without the compressors) used in semiconductor manufacturing [13, 140].

Cryopumps are still the most popular applications of cryocoolers and currently about 20,000 cryopumps are manufactured per year [3].

1.8.2 MRI Systems

With the advent of MRI technology, G-M and G-PTCs have found themselves a huge market in clinical diagnostics, medical research and NMR spectroscopy devices. Most MRI scanners use 10 K G-M coolers for shield cooling. Due to development of 4 K G-M coolers, recondensing zero boil-off (ZBO) cryostats for MRI have been available for the past 5–6 years. Figure 1.25 shows the growth in sales of MRI machines in the past few years [3].

It may be noticed that there has been a 20% increase in MRI sales in the past 4 years, which highlights the technological advancement and huge demands of modern MRI systems.

All commercial MRI systems available in hospitals and clinics all around the world use low-temperature superconducting (LTS) magnets. Therefore, since they require temperatures of 4 K level, they typically use a liquid helium bath to maintain the magnet in superconducting state. Hence, different configurations are possible for an MRI machine to fulfil this requirement. Table 1.1 shows the developments in MRI systems over the years [141].

It may be noticed from this table that the MRI machines mainly use G-M cryocoolers. With the development of 4 K G-M coolers, their integration with MRI magnets resulted in the development of zero boil-off (ZBO) technology, which could

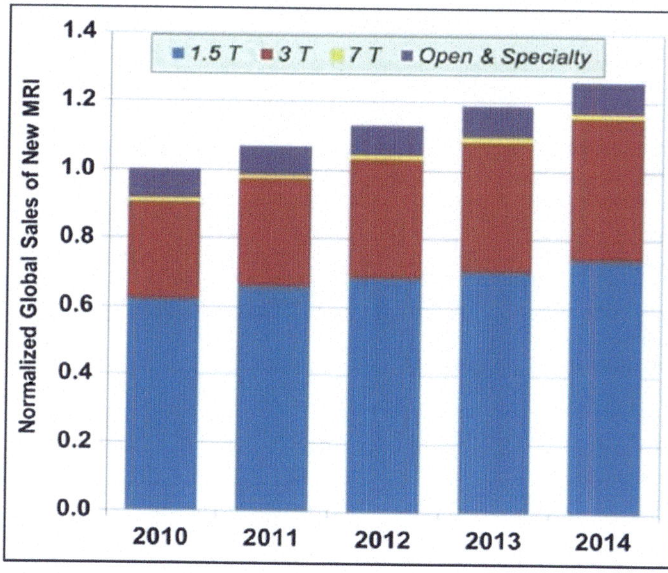

Fig. 1.25 Worldwide sales growth in MRI [3]

recondense the LHe vapours at 4 K, whereas a single shield could be conductively cooled by the first stage of the G-M cooler [4]. A schematic of such a system is shown in Fig. 1.26.

The ZBO technology has helped to reduce the costs associated with the periodic LHe refills, thus increasing the affordability of MRI for hospitals and clinics with minimized down-time. Most modern commercial MRI systems of 1.5–3 T utilize this ZBO technology. MRI systems typically use G-M cryocoolers. However, due to displacer and rubbing seal maintenance requirement, vibrations and EMI caused, manufacturers are looking to replace the G-M coolers with GM-PTCs for the ZBO systems.

In addition, the latest advancements in MRI technology are motivated towards cryogen-free open MRI systems, where the SC magnets are conductively cooled by the cryocooler. This methodology has potential to eliminate the large LHe tanks in the MRI system, thereby reducing its size requirements and associated maintenance. However, maintaining a large homogeneous magnetic field for high-quality images poses a problem for their acceptance in the market. The current research efforts in MRI technology are being made towards invisible cryogenics [142], thereby eliminating the end user interaction with the inherent cryogenic equipment involved in the MRI system.

Table 1.1 Developments in MRI Systems[141]

Sr no.	Cooling devices or medium	Temperature		MRI magnet cooling mode
		Shield	Magnet	
1	Magnet immersed in LHe bath surrounded by LN_2 shield	80	4.2	Immersion in LHe
2	Magnet immersed in LHe bath surrounded by two conductively cooled shields using two-stage G-M cryocooler	80–20	4.2	Immersion in LHe
3	Magnet and one thermal shield conductively cooled by two-stage G-M cryocooler	50	10	Conduction cooled
4	Magnet immersed in a recondensing LHe bath surrounded by one conductively cooled shield using a two-stage G-M cryocooler	50	4.2	Immersion in LHe

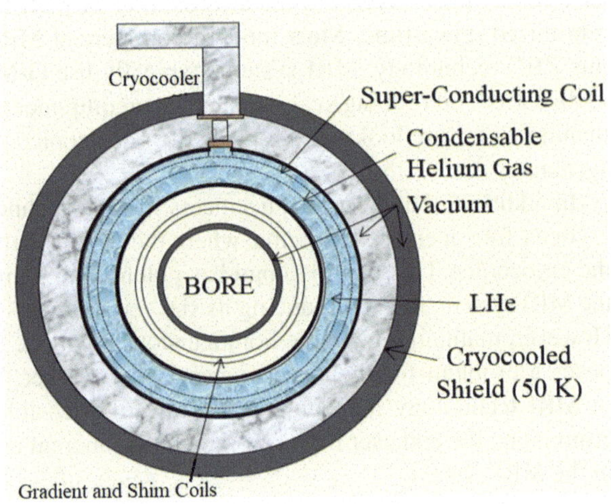

Fig. 1.26 Zero boil-off recondensing MRI system

1.8.3 NMR

Nuclear magnetic resonance (NMR) spectroscopy is a technique that is based on the study of interaction of electromagnetic radiation with matter. In recent years, no other technique has gained such significance as NMR spectroscopy. It is used in all branches of science in which precise structural determination is required and in which the nature of interactions and reactions in solution is being studied.

NMRs are the predecessors of the modern full-body MRI scanners and have similar cooling requirements. They are used to study the physical, chemical and biological properties of matter [143]. However, they require 20 very high magnetic fields (10–25 T) with higher magnetic field homogenity as compared to MRI scanners. In addition, the vibration sensitivity requirements are extremely stringent in order to obtain a high signal-to-noise ratio (SNR). Hence, most NMRs prefer to use LHe baths as compared to only cryocoolers.

However, the operating and maintenance costs in this case are extremely high. Due to rising helium costs and cryogen replenishment demands, many researchers and users are working to develop cryocooler-based NMRs due to their improved reliability and lower operating costs. NMR cryostats have been developed and investigated using cryocoolers by many researchers for HTS and LTS magnet-based NMRs [144–147].

Figure 1.27 shows a schematic diagram and demonstration model of the first developed actively cooled 400 MHz NMR magnet cryostat with ZBO technology using a 4 K GM-PTC.

In addition, several models of cryogenically cooled probes have been developed [148–151], which indirectly cool the electronics of the probe and increase the SNR. Cryogenically cooled probes typically employ a G-M or GM-PTC to cool the NMR probe.

Figure 1.28 shows a cryogenic cooling circuit for an NMR cold probe, developed by Oxford Instruments® Superconductivity [151].

This system uses a PT810 from Cryomech Inc.®, which gives a cooling effect of 15 W at 15 K on the second stage with a power input of 7.5 kW.

This cooling circuit can provide a refrigeration effect of about 3.5 W at 25 K to the RF coils in the probe and 2.5 W to the electronics [151].

Although several technology demonstration models exist, the NMRs with ZBO cryostats are still in the research phase. In addition, several models of conductively cooled HTS systems are also an emerging area for researchers. However, the ZBO technology has been successfully incorporated in commercial MRI systems. In addition, cryogenic probes for NMR applications are commercially available.

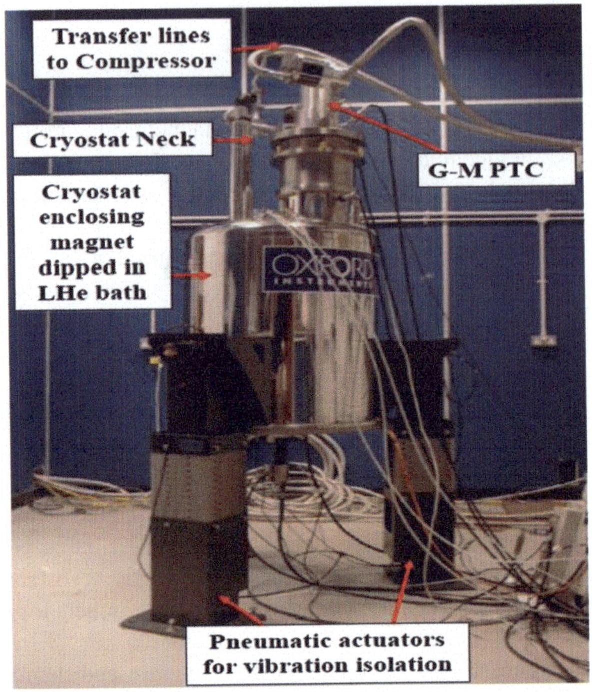

Fig. 1.27 Developed model of an actively cooled 400 MHz NMR cryostat [147]

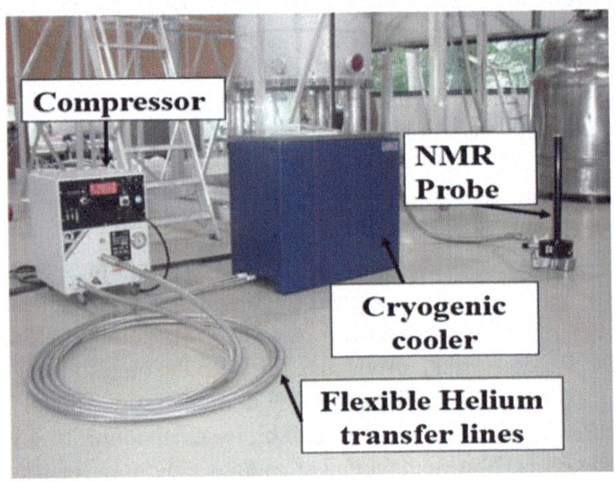

Fig. 1.28 Cryogenic cooling circuit for NMR cold probe developed at Oxford Instruments® Superconductivity [151]

1.8.4 IR Sensors/Detectors

Along with cryopumps, cooling of IR detectors to temperatures of about 80 K for space and military applications are one of the earliest applications of cryocoolers. It still remains a huge market for the Stirling cryocoolers [152]. Due to the requirement of fast cool-down time for some applications, open cycle J-T coolers are deployed for IR seekers. These J-T coolers are usually of the single-fluid type having nitrogen (N_2) or argon (Ar) as the working fluid. The gas is charged at 200–400 bar pressures, while the cool-down time is usually less than a minute. Stirling coolers used for the cooling of IR sensors typically produce 0.25–1.5 W of cooling effect at 80 K with a power input of 40–150 W. The use of such cryocoolers in different configurations has been investigated and developed by many researchers [152–154]. Stirling and J-T coolers are usually preferred for IR surveillance detectors since they can be easily miniaturized and have a high cooling effect. However, the vibrations in the Stirling cooler due to the displacer movement may affect the SNR signal.

Development of cryocoolers for space and military applications has led to improved reliability and lifetimes. Stirling cryocoolers have a huge market in space applications due to higher COPs as compared to other types of cryocoolers. Due to the problems associated with SNR, a lot of research work is being carried out to replace the Stirling cryocooler with Stirling-type PTCs [155–158]. However, the orientation dependency of PTCs may prove to be a major challenge for its adaptability for spacecraft IR detector cooling.

1.8.5 Dry Dilution Refrigerator (DR)

A closed cycle cryogen-free He^3/He^4 dilution refrigerator is by far the most efficient continuous refrigeration methods to reach temperatures in the millikelvin range (below 0.3 K) [159]. The ultra-low temperatures provided by DRs find applications in many fields like solid state physics, materials research, nuclear physics, neutron-scattering or astrophysics and quantum information technology [160, 161].

Such cryogen-free dilution refrigerators use G-M coolers or G-PTCs instead of LN_2 and LHe bath to precool the system. In this way, they allow for a bigger space in the mixing chamber and are easy to operate. In addition, it can be completely computer controlled. Use of G-PTCs in these systems also allows for a low vibration operation.

Figure 1.29 shows a schematic diagram of an experimental DR with a separate 1 K loop [161]. DRs are basically commercially built cryostats. However, in some special cases, researchers do construct their own DRs and conduct studies on its performance [160–166].

Figure 1.30 shows an Oxford Instruments® make Triton cryogen-free dry dilution refrigerator.

This dilution refrigerator uses a Cryomech® GM-PTC and delivers 400 W cooling power at 100 mK and can attain low temperatures of about 10 mK [168].

1.8.6 Cryocooler-Based Liquefiers

Small-scale nitrogen liquefiers using multi-cylinder Stirling cryogenerators are quite popular for laboratory and small industry applications. These Stirling cryocoolers

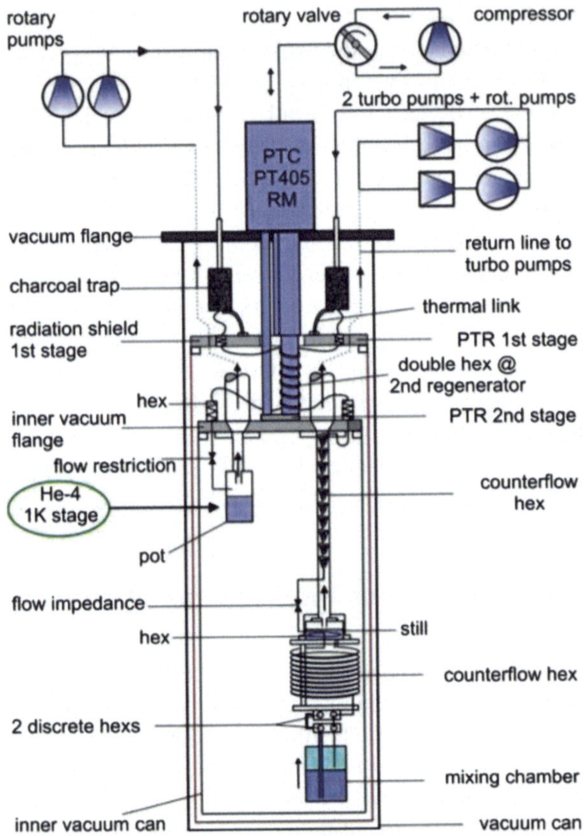

Fig. 1.29 Cross section of an experimental DR unit with a separate 1 K-stage [167]

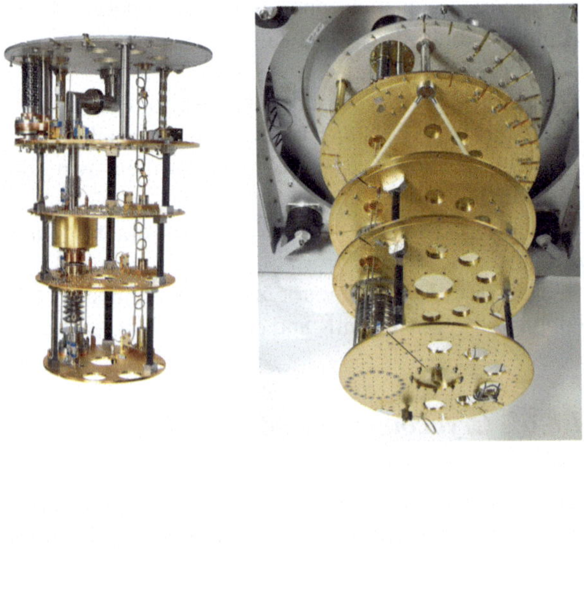

Fig. 1.30 Triton cryogen-free dilution refrigerator by Oxford Instruments® [168]

Fig. 1.31 Stirling® cryogenerator-based nitrogen liquefier [169]

generally use oil-flooded compressors. The liquefaction rate for such nitrogen liquefiers ranges from 5 to 100 lph depending on the number of cylinders used for liquefaction. Figure 1.31 shows a LN_2 liquefier which has a liquefaction rate of about 90 lph with a power input of 129 kW [169].

G-M cooler-based nitrogen liquefiers are available for small liquefaction rates. G-PTC-based helium liquefiers were developed and commercialized by Cryomech® [170–174], mainly for small laboratory applications. The nitrogen liquefiers by Cryomech® use one or two single-stage G-M coolers to deliver a liquefaction rate between 1 and 10 lph with a power input of 10–25 kW. Similarly, the helium liquefier models uses one to three G-PTCs to deliver a liquefaction rate of 0.25–2.5 lph with a power input of 10–30 kW [174]. Although the liquefaction rates for such liquefiers is quite low, these are found suitable for small laboratories in remote locations.

1.8.7 Miniaturized J-T Cryocooler

Miniaturized J-T cryocooler or microminiature refrigerators (MMRs) was conceptualized, developed and commercialized by W. A. Little of MMR Technologies Inc. in the early 1980s [175]. Although the J-T cycle is less efficient than other alternate cycles, the absence of mechanical moving parts makes it an ideal choice for micro-miniaturization. Such MMRs have potential applications in small cryogenic devices that dissipate very little heat during operation and hence require few microwatts to a few milliwatts of cooling effect. These MMRs are fabricated using a proprietary and patented photolithographic process on glass substrates. A typical N_2 microminiature J-T cryocooler fabricated using this technique delivers a cooling effect of about 250 mW at 80 K and the typical dimensions of its cold stage is 6 cm × 1.4 cm × 0.2 cm. Figure 1.32 shows the MMRs developed by MMR Technologies Inc. [176].

In addition to MMRs developed by photolithographic techniques, micromachined miniaturized J-T cryocoolers have been developed by University of Twente

Fig. 1.32 Microminiature J-T refrigerators developed by MMR Technologies Inc. [176]

[177]. The cold stage of these coolers consists of a stack of three micromachined glass plates. The high and low-pressure lines are etched as rectangular channels on the top and bottom plate or wafer with supporting pillars. These wafers are plated with a thin layer of gold on their outer surface to minimize parasitic radiation heat loss [178].

As these are open cycle systems, they are operated with nitrogen gas with pressures ranging between 6 and 100 bar. The cooling effect obtained from these coolers is about 10–20 mW at 100 K. The typical dimensions of a cold stage with a net cooling power of 10 mW at 100 K are 30 × 2.2 × 0.7 in mm. Micromachined cryocoolers are attractive tools for cooling electronic chips and devices to cryogenic temperatures. Recently, a two-stage 30 K microcooler operating with nitrogen and hydrogen gas was developed at the University of Twente using micromachining processes. The cooling powers typically range from 21 to 84 mW at 95 K at the nitrogen stage, corresponding to 30–5 mW at 31–32 K at the hydrogen stage [179].

The current research in micro-miniature J-T cryocoolers is focused on developing units with higher net cooling powers (up to 200 mW). In addition, elimination of water clogging, use of sorption compressors and development of multistage microcoolers for specialized applications like low noise amplifiers and optical detectors, are currently being investigated [180–184].

1.9 Invisible and Reliable Cryogenics

Tremendous improvement in cryocooler technology has taken place since the 1960s after the commercialization of Stirling cryocoolers. Spacecraft technology, helium and nitrogen liquefaction, MRI and NMR spectroscopy, are all high-demand applications which have been the driving factors for the demand for better efficiencies and longer lifetimes for cryocoolers in the past two decades. The trends in cryocoolers that have occurred in the recent years have been described in this chapter. By virtue of these advancements, several new applications for cryocoolers are being explored in various research laboratories, industries and universities. In addition, the development for better and more efficient regenerator materials is also ongoing. At the same time, due to higher reliability of cryocoolers, cryogenics is becoming invisible to users. With the rising costs of cryogens, particularly helium, cryocooler technology will continue to attract major funding and resources for further development in this area.

References

1. Walker G (1983) Cryocoolers part—1 & 2. Plenum Press, New York
2. PTI, London (2010) Earth's helium reserves to run out by 2030', Asian Age—23rd August
3. Lvovsky Y, Stautner EW, Zhang T (2013) Novel technologies and configurations of superconducting magnets for MRI. Supercond Sci Technol 26:093001
4. Radebaugh R (2009) Cryocoolers: the state of the art and recent developments. J Phys Condens Matter 21(16):164219
5. Atrey MD (2012) Cryogenic Engineering NPTEL Online Lectures
6. Mehta RN, Bapat SL, Atrey MD (2012) Characterization of sorption compressor for mixed refrigerant J-T cryocooler. AIP Conf Proc 1434:1797–1804
7. Jha AR (2006) Cryogenic Technology and Applications. Butterworth-Heinemann, Oxford, p 187
8. Kaka S, Avelino MR, Smirnov HF (2003) Low temperature and cryogenic refrigeration, NATO science series II, vol 99. Springer Science & Business Media, New York, p 420
9. Bapat SL (2000) Performance estimates of Stirling cycle cryocooler using two-component two-phase mixture. In: Proceedings of ICEC-18, Mumbai, pp 607–610
10. AIM INFRAROT-MODULE Gmbh product catalogue
11. Davey G (1990) Review of the Oxford cryocooler. In: Advances in cryogenic engineering, vol 35. Springer, Boston
12. Trevisani L, Kuriyama T, Negrini F, Okamura T, Ohtani Y, Okamura M, Fabbri M (2002) Performance improvement of a two-stage Gifford-McMahon cryocooler with Er(Ni0.075Co0.925)2 magnetic regenerator material. Cryogenics 42:653–657
13. Sumitomo Heavy Industries Ltd. Cryocooler product catalogue
14. Gifford WE, Longsworth RC (1963) Pulse tube refrigeration process. In: Advances in cryogenic engineering, vol 10B. Springer, Boston, pp 69–78
15. Tarasov AA, Shkrebyonock MP, Mikulin EI (1984) Low temperature expansion pulse tube. In: Advances in cryogenic engineering, vol 29. Springer, Boston, pp 629–637
16. Zhu S, Wu P, Chen Z (1990) Double inlet pulse tube refrigerators: an important improvement. Cryogenics 30:514–520
17. Gao JL, Matsubara Y (1994) Experimental investigation of 4 K pulse tube refrigerator. Cryogenics 34:25–30
18. Thummes G, Bender S, Heiden C (1996) Approaching the 4He lambda-line with a liquid nitrogen precooled two-stage pulse tube refrigerator. Cryogenics 36:709
19. Thummus G, Wang C, Heiden C, (1998) Small scale 4He liquefaction using a 4 K pulse tube cryocooler. Advances in cryogenic engineering, Springer, Boston 38 3 337-342.
20. Wang C (2001) Helium liquefaction with a 4K pulse tube cryocooler. Cryogenics 41:491–496
21. Wang C (2009) Small scale helium liquefaction systems. J Phys Conf Ser 150:012053
22. Yang LW, Thummes G (2005) High frequency two stage pulse tube refrigerator with base temperature below 20 K. Cryogenics 45:155–159

23. Gardner DL, Swift GW (1997) Use of inertance in orifice pulse tube refrigerators. Cryogenics 37:117–121
24. Radebaugh R (1990) A review of pulse tube refrigeration. In: Advances in cryogenic engineering, vol 35. Springer, Boston, pp 1191–1205
25. Matsubara Y (1998) Classification of pulse tube cryocoolers. In: International cryogenic engineering conference, vol 17, pp 11–16
26. https://trc.nist.gov/cryogenics/
27. Bailey PB, Dadd MW, Stone CR (2012) High speed compressors, cryocoolers 17. ICC Press, Boulder, pp 347–356
28. Davies GF, Eames IW, Bailey PB, Dadd MW, Janiszewski A, Stone CR, Maidment GG, Agnew B (2010) Cooling microprocessors using vapor compression refrigeration. In: IEEE Itherm Conference, pp 1–8. Abstract
29. Davies G, Eames I, Bailey P, Dadd M, Janiszewski A, Stone C, Maidment G, Agnew B (2009) Development of a miniature vapor compression refrigeration system for electronic cooling. In: Proceedings of the ASME InterPack Conference, San Francisco, CA, USA, vol 2, pp 399–408. Abstract
30. Cheuk CF, Hill NG, Strauch R, Bailey PB, Raab J (2003) Producibility of cryocooler compressors. In: Cryocoolers, vol 12. Kluwer Academic/Plenum Press, New York, pp 275–281
31. Dadd MW, Bailey PB, Davey G, Davis T, Thomlinson BJ (2002) Vibration reduction in balanced linear compressors. In: Cryocoolers, vol 11. Plenum Press, New York, pp 175–182
32. Davey G (1990) Review of the Oxford cryocooler. In: Advances in cryogenic engineering, vol 35B. Springer, Boston, pp 1423–1430
33. Orlowska AH, Davey G (1987) Measurement of losses in a Stirling cycle cooler. Cryogenics 27:645–651
34. Benschop T, Mullie J, Bruins P, Martin JY (2003) Development of 6-W high reliability cryocoolers at Thales. Cryogenics ICEC-19:33–42
35. QDrive®product catalogue
36. Technical information. Kryolab, Lund University. Retrieved 26 Jan 2013
37. Keppler F, Nellis G, Klein SA (2004) Optimization of the composition of a gas mixture in a Joule-Thomson cycle. HVAC&R Res 10(2):213–230
38. Ravishankar N, Ardhapurkar PM, Atrey MD (2014) Prediction of differential Joule-Thomson inversion curves for cryogens using equations of state, NSC-25, Hyderabad
39. Podbiclnak WJ (1936) US patent specification 2,041,725
40. Little WS (2008) Microminiature refrigeration. In: Advances in cryogenic engineering, vol 53. Springer, Boston, pp 597–605
41. Alexeev A, Haberstroh C, Quack H (1998) Further development of a mixed gas Joule Thomson refrigerator. In: Advances in cryogenic engineering, vol 43. Springer, Boston, pp 1667–1674
42. deWaele ATAM (2011) Basic operation of cryocoolers and related thermal machines. J Low Temp Phys 164:179–236
43. Atrey MD, Bapat SL, Narayankhedkar KG (1990) Cyclic simulation of Stirling cryocoolers. Cryogenics 30(4):341–347
44. Atrey MD, Bapat SL, Narayankhedkar KG (1991) Theoretical analysis and performance investigation of Stirling cycle regenerators. Cryogenics 31(12):1044–1052
45. Atrey MD, Bapat SL, Narayankhedkar KG (1988) An approach to the analysis of Stirling cycle Cryocoolers regenerator using finite difference techniques. In: Proceedings of 2nd international conference on cryogenics. Kolkata, India, pp 305–315
46. Atrey MD, Bapat SL, Narayankhedkar KG (1993) Optimization of design parameters of Stirling cycle machine. Cryogenics 33(10):951–957
47. Atrey MD, Heiden C (1996) Performance evaluation of an optimized two-stage, free-displacer plastic Stirling cryocooler with gap regenerator. Cryogenics 36(1):47–52
48. Atrey MD, Bapat SL, Heiden C (1994) Development of a computer model for three-stage, split type, free displacer Stirling cryocooler. Cryogenics 34(9):727–732

49. Atrey MD, Bapat SL, Heiden C (1994) Optimisation of design parameters of two stage, split type, free displacer Stirling cryocooler. Cryogenics 34:215–218
50. Atrey MD, Bapat SL, Narayankhedkar KG (1994) Influence of regenerator matrix and working fluid on optimisation of design parameters of Stirling cryocoolers. Cryogenics 34:211–214
51. Walker G, Weiss M, Fauvel R, Reader G (1989) Microcomputer simulation of Stirling cryocoolers. Cryogenics 29(8):846–849
52. Zhang C-q, Zhong C (2015) Theoretical modeling of a gas clearance phase regulation mechanism for a pneumatically-driven split-Stirling-cycle cryocooler. Cryogenics 66:13–23
53. Chen X, Wu YN, Zhang H, Chen N (2009) Study on the phase shift characteristic of the pneumatic Stirling cryocooler. Cryogenics 49(3–4):120–132
54. Park SJ, Hong YJ, Kim HB, Koh DY, Kim JH, Yu BK, Lee KB (2002) The effect of operating parameters in the Stirling cryocooler. Cryogenics 42(6–7):419–425
55. Organ AJ (1999) The miniature, reversed Stirling cycle cryo-cooler: integrated simulation of performance. Cryogenics 39(3):253–266
56. Deac IG (1994) Design and performance test of a miniature Stirling cryocooler. Cryogenics 34:191–193
57. Bauwens L (1994) Adiabatic losses in Stirling cryocoolers: a stratified flow model. Cryogenics 34(8):627–633
58. Razani A, Dodson C, Roberts T A model for exergy analysis and thermodynamic bounds of Stirling refrigerators. Cryogenics 50(4):231–238
59. Tailor PR, Narayankhedkar KG (1988) Analysis and performance prediction of electromagnetically-driven free displacer Stirling cryocooler. Cryogenics 28(3):169–176
60. Walker G (1983) Design guidelines for large Stirling cryocoolers. Cryogenics 23(2):113–114
61. He Y-L, Zhang D-W, Yang W-W, Gao F (2014) Numerical analysis on performance and contaminated failures of the miniature split Stirling cryocooler. Cryogenics 59:12–22
62. Li R, Grosu L (2017) Parameter effect analysis for a Stirling cryocooler. Int J Refrig 80:92–105
63. Yang X, Chung JN (2005) Size effects on miniature Stirling cycle cryocoolers. Cryogenics 45(8):537–545
64. Xia M, Chen X (2010) Analysis of resonant frequency of moving magnet linear compressor of Stirling cryocooler. Int J Refrig 33(4):739–744
65. McMahon HO, Gifford WE (1960) A new low temperature gas expansion cycle part I. In: Advances in cryogenic engineering, vol 5. Springer, Boston, pp 354–366
66. Gifford WE (1966) The Gifford McMahon cycle. In: Advances in cryogenic engineering, vol 11. Springer, Boston, pp 152–159
67. Ackermann RA, Gifford WE (1971) A heat balance analysis for Gifford McMahon cryorefrigerator. In: Advances in cryogenic engineering, vol 16. Springer, Boston, pp 221–229
68. Herycak P (1963) Thermodynamic analysis of a new gas refrigeration cycle. Cryogenics 3:23–26
69. Tirumaleshwar M, Subramanyam SV (1986) Heat balance analysis of single stage Gifford McMahon cycle cryorefrigerator. Cryogenics 26:189–195
70. Tirumaleshwar M, Subramanyam SV (1986) Gifford McMahon Cycle—a theoretical analysis. Cryogenics 26:177–188
71. Tirumaleshwar M, Subramanyam SV (1986) Two stage Gifford McMahon cycle cryorefrigerator operating at 20 K. Cryogenics 26:547–555
72. Minas C, Hualde PM (1992) Dynamic modelling of a Gifford McMahon cryorefrigerator. Cryogenics 32:634–639
73. Huang W, Wu PY, Hu SL, Zhang L, Zhout YM, Gong LH, Chen GB (1996) Dynamic simulation of one-stage G-M refrigerator and comparison with experiment. Cryogenics 36(9):643–647
74. Atrey MD, Hfner HU, Heiden C (1996) Development of a computer model for two stage G-M cryocooler. In: Advances in cryogenic engineering, vol 41A. Springer, Boston, pp 1593–1600

75. Duttagupta B, Kush PK, Atrey MD (2000) Cyclic analysis and experimental investigation of a single stage G-M cryocooler. In: Proceedings of ICEC-18, Mumbai, pp 579–583
76. Gifford WE, Longsworth RC (1963) Pulse tube refrigeration ASME. J Eng Ind 86(3):264–268
77. Narayankhedkar KG, Mane VD (1973) Investigation of pulse tube refrigerator. J Eng Ind 95(1):373–378
78. Longsworth RC (1967) An experimental investigation of pulse tube refrigeration heat pumping rates. In: Advances in cryogenic engineering, vol 12. Springer, Boston, pp 608–618
79. Radebaugh R (2003) Thermodynamics of regenerative refrigerators. Generation of low temperature and its application. Technical Center, Kamakura, pp 1–20
80. Lokanath M, Atrey MD (2010) Phasor analysis of pulse tube refrigerator. In: International cryocooler conference, vol 16, pp 299–308
81. Chen N, Yang CG, Xu L (2005) Optimal vector analysis of the phase shifter in the pulse tube refrigerator. In: Proceedings of the twentieth international cryogenic engineering conference (ICEC20), pp 261–264
82. Zhu SW, Chen ZQ (1994) Isothermal model of pulse tube refrigerator. Cryogenics 34:591–595
83. Atrey MD, Narayankhedkar KG (2000) Development of second order isothermal model of the orifice type pulse tube refrigerator. In: ICEC 18, pp 519–522
84. Yang LW (2002) Shuttle loss in pulse tubes. Cryocooler 11:353–362
85. Gawali BS, Atrey MD, Narayankhedkar KG (2002) Performance prediction and experimental investigation on orifice PTC. In: International cryognic engineering conference, vol 19, pp 391–395
86. Gawali BS, Narayankhedkar KG (2006) Performance prediction and experimental investigation on integral pulse tube refrigerator for 15 W at 70 K using indigenously developed linear compressor. In: Advances in cryogenic engineering, vol 51. Springer, Boston, pp 11–18
87. Zhu S, Matsubara Y (2004) Numerical method of inertance tube pulse tube refrigerator. Cryogenics 44:649–660
88. Wu P, Wang C, Chen Z (1992) Numerical modeling of an orifice pulse tube refrigerator. Cryogenics 32:785–790
89. Wu P, Wang C, Chen Z (1993) Numerical modeling of an double inlet pulse tube refrigerator. Cryogenics 33:526–530
90. Zhou Y, Ju YL, Wang C (1998) Numerical simulation and experimental verification of oscillating flow in the pulse tube refrigerator. Cryogenics 38:169–176
91. Xu M, Chen Z, He Y, Gao C, Tao W (2001) Numerical simulation of convergent and divergent tapered pulse tube refrigerators and experimental verification. Cryogenics 41:699–704
92. Patankar SV, Spalding DB (1972) A calculation procedure for heat, mass and momentum transfer in three-dimensional parabolic flows. Int J Heat Mass Transfer 15(10):1787–1806
93. Desai PV, Harvey JP, Kirkconnell CS, Cha JS, Ghiaasiaan SM (2006) Multidimensional flow effects in pulse tube refrigerators. Cryogenics 46:658–665
94. Barrett F, Razani A (2004) Modeling pulse tube cryocooler with CFD. In: Advances in cryogenic engineering, vol 49. Springer, Boston, pp 1493–1499
95. Zhang XB, Qiu LM, Gan ZH, He YL (2007) CFD study of a simple orifice pulse tube cooler. Cryogenics 47:315–321
96. He YL, Tao YB, Gao F (2009) A new computational model for entire pulse tube refrigerators: model description and numerical validation. Cryogenics 49:84–93
97. Kamoshita T, Yasukawa Y, Ohshima K (2002) Miniature pulse tube cooler. Fuji Electr J 75(5)
98. Nachman I, Pundak N, Grossman G (2009) CFD modeling of reciprocating flow around a bend in pulse tube cryocoolers. Cryocooler 1:251–259
99. Liang W, de Waele ATAM (2007) A new type of streaming in pulse tubes. Cryogenics 47:468–473
100. Du BY, Yang LW, Cai JH, Liang JT (2007) Numerical simulation of a regenerator in a two-stage pulse tube refrigerator. Cryocoolers 14:405–409

101. Banjare YP, Sahoo RK, Sarangi SK (2009) CFD simulation of a Gifford-Mcmahon type pulse tube refrigerator. Int J Therm Sci 4:1–8
102. Gedeon D (2011) Sage v9 edition—user guide Stirling, pulse-tube and low-TCooler Model Classes. Gedeon Associates, Athens
103. Gedeon D (1994) Sage: object orientated software for Stirling-type machine design, vol 4. American Institute for Aeronautics and Astronautics, pp 1902–1907
104. Peyret RP, Taylor TD (1983) Computational methods for fluid flow. Springer, New York
105. Cao Q, Gan ZH, Liu GJ, Li ZP, Wu YZ, Qui LM, Pfotenauer JM (2009) Theoretical and experimental study on a pulse tube cryocooler driven with a linear compressor. Cryocooler 15:149–156
106. Gedeon DR, Wilson KB (2004) Status of pulse tube refrigerator development at sunpower. Cryocoolers 13:31–35
107. Emery N, Caughley A, Glasson N, Tucker A, Gschwendtner M (2011) Development of high frequency pulse tube. Cryocooler 16:75–82
108. Maytal BZ (1994) Performance of ideal flow regulated Joule-Thomson cryocooler. Cryogenics 34:723–726
109. Xue H, Ng KC, Wang JB (2001) Performance evaluation of the recuperative heat exchanger in on a miniature Joule-Thomson cooler. Appl Therm Eng 21:1829–1844
110. Ng KC, Xue H, Wang JB (2002) Experimental and numerical study on a miniature Joule-Thomson cooler for steady-state characteristics. Int J Heat Mass Transfer 45:609–618
111. Hong YJ, Park SJ, Choi YD (2009) A numerical study of the performance of a heat exchanger for a miniature Joule-Thomson refrigerator. In: International cryocooler conference cryocoolers 15, pp 379–386
112. Ardhapurkar PM, Atrey MD (2014) Performance optimization of a miniature Joule-Thomson cryocooler using numerical model. Cryogenics 63:94–101
113. Chou FC, Pai CF, Chien SB, Chen JS (1995) Preliminary experimental and numerical study of transient characteristics for Joule-Thomson cryocooler. Cryogenics 35:311–316
114. Chien SB, Chen JS, Chou FC (1996) A study on the transient characteristics of a selfregulating Joule-Thomson cryocooler. Cryogenics 36:979–984
115. Damle RM, Atrey MD (2015) Transient simulation of a miniature Joule-Thomson (J-T) cryocooler with and without the distributed J-T effect. Cryogenics 65:49–58
116. Gong MQ, Luo EC, Wu JF, Zhou Y (2002) On the temperature distribution in the counter flow heat exchanger with multicomponent non-azeotropic mixtures. Cryogenics 42:795–804
117. Ardhapurkar PM (2015) Investigations on two-phase heat exchanger for mixed refrigerant Joule-Thomson cryocooler. PhD thesis, IIT Bombay
118. Venkatatrathnam G (2008) Cryogenic mixed refrigerant processes. Springer, New York
119. Walimbe NS, Narayankhedkar KG, Atrey MD (2010) Experimental investigation on mixed refrigerant Joule-Thomson cryocooler with flammable and non-flammable refrigerant mixtures. Cryogenics 50(10):653–659
120. Little WA (1997) Method for efficient counter-current heat exchange using optimized mixtures. US Patent 5,644,502
121. Gong MQ, Luo EC, Zhou Y, Liang JT, Zhang L (2000) Optimum composition calculation for multicomponent cryogenic mixture used in Joule-Thomson refrigerators. In: Advances in cryogenic engineering, vol 45. Springer, Boston, pp 283–290
122. Alexeev A, Quack H (2003) Refrigerant mixture for a mixture-throttling process. US Patent 6,513,338 B1
123. Radebaugh R (1995) Recent developments in cryocoolers. In: Proceedings of the 19th international congress of refrigeration, vol III(B), pp 973–989
124. Alexeev A, Thiel A, Haberstroh C, Quack H (2000) Study of behavior in the heat exchanger of a mixed gas Joule-Thomson cooler. In: Advances in cryogenic engineering, vol 45. Springer, Boston, pp 307–314
125. Gong MQ, Wu JF, Luo EC, Qi YF, Hu QG, Zhou Y (2002) Study on the over-all heat transfer coefficient for the tube-in-tube heat exchanger used in mixed-gases coolers. In: Advances in cryogenic engineering, vol 47B. Springer, Boston, pp 1483–1490

126. Ardhapurkar PM, Sridharan A, Atrey MD (2012) Investigations on two-phase heat exchanger for mixed refrigerant Joule-Thomson cryocooler. In: Advances in cryogenic engineering: transactions of the cryogenic engineering conference—CEC, vol 57. AIP Publishing, pp 706–713
127. Ardhapurkar PM, Sridharan A, Atrey MD (2014) Experimental investigation on temperature profile and pressure drop in two-phase heat exchanger for mixed refrigerant Joule-Thomson cryocooler. Appl Therm Eng 66:94–103
128. Gong M, Ma Jia WJ, Yu Z, Zhaohu S, Zhou Y (2009) Nucleate pool boiling of liquid methane and its natural gas mixtures. Int J Heat Mass Transfer 52:2733–2739
129. Nellis G, Hughes C, Pfotenhauer J (2005) Heat transfer coefficient measurements for mixed gas working fluids at cryogenic temperatures. Cryogenics 45(8):546–556
130. Little WA (2008) Heat transfer efficiency of Kleemenko cycle heat exchangers. In: Advances in cryogenic engineering, vol 53. Springer, Boston, pp 606–613
131. Silver L (1947) Gas cooling with aqueous condensation. Trans Inst Chem Eng 25:30–42
132. Bell KJ, Ghaly MA (1973) An approximate generalized design method for multicomponent/partial condenser. In: AIChE symposium series, vol 69, pp 72–79
133. Ardhapurkar PM, Sridharan A, Atrey MD (2014) Flow boiling heat transfer coefficients at cryogenic temperatures for multi-component refrigerant mixtures of nitrogen-hydrocarbons. Cryogenics 59:84–92
134. Gungor KE, Winterton RHS (1987) Simplified general correlation for saturated flow boiling and comparisons with data. Can J Chem Eng 65(1):148–156
135. Granryd E (1991) Heat transfer in flow evaporation of non-azeotropic refrigerant mixtures—a theoretical approach. In: Proceedings of the 18th international congress of refrigeration, Montreal, vol 3, pp 1330–1334
136. Ardhapurkar PM, Sridharan A, Atrey MD (2014) Performance evaluation of heat exchanger for mixed refrigerant J-T cryocooler. Cryogenics 63:49–56
137. Cavallini A, Zecchin RA (1974) A dimensionless correlation for heat transfer in forced convection condensation. In: 6th international heat transfer conference, Tokoyo, vol 3, pp 309–313
138. Radebaugh R (2004) Refrigeration for superconductors. Proc IEEE 92:1719–1734
139. Day C (2006) Basics and applications of cryopumps CERN accelerator school proceedings on vacuum in accelerators
140. Oerlikon Leybold Vacuum GmbH-Product catalogue
141. Ackermann RA, Herd KG, Chen WE (1999) Advanced cryocooler cooling for MRI systems. In: Cryocoolers 10. Kluwer Academic/Plenum Publishers, New York, pp 857–867
142. Steinmeyer F, Retz PW, White K, Lang A, Stautner W, Smith PN, Gilgrass G (2002) Towards the invisible cryogenic system for magnetic resonance imaging. In: Advances in cryogenic engineering, vol 47B, AIP conference proceedings, vol 613, pp 1659–1666
143. Webb GA (2013) Annual reports on NMR spectroscopy. Academic Press, New York
144. Yoon S, Cheon K, Lee H, Moon S-H, Kim S-Y, Kim Y, Park S-H, Choi K, Hong G-W (2013) The performance of the conduction cooled 2G HTS magnet wound without turn to turn insulation generating 4.1 T in 102 mm bore. Phys C: Superconduct 494:242–245
145. Bae JH, Kim SH, Kim HJ, Sohn MH, Seong KC, Kim HM (2009) Design, fabrication and evaluation of a conduction cooled HTS magnet for SMES. Phys C Superconduct 469(15–20):1794–1798
146. Tanaka H, Fukuda Y, Okada M, Saho N, Saitoh K, Kitaguchi H (2008) Development of cryogenic cooling system using a double G-M cryocooler for NMR spectrometer. J Phys Conf Ser 97:012232
147. Kirichek O, Carr P, Johnson C, Atrey M (2005) Nuclear magnetic resonance magnet actively cooled by pulse tube refrigerator. Rev Sci Instrum 76:055104. https://doi.org/10.1063/1.1896949
148. Kovacsa H, Moskaua D, Spraulb M (2005) Cryogenically cooled probes a leap in NMR technology. Prog Nucl Magn Reson Spectrosc 46:131–155

149. Kotsubo V, Nast R (1996) Cryogenic system for a high temperature superconductor NMR probe. In: Advances in cryogenic engineering, vol 41A. Springer, Boston, pp 1857–1864
150. Saitoh K, Yamamoto H, Kawasaki K, Fukuda Y, Tanaka H, Okada M, Kitaguchi H (2008) Development of cryogenic probe system for high-sensitive NMR spectroscopy. J Phys Conf Ser 97:012141
151. Atrey MD, Healy A, Janaway T, McMahon A (2005) Pulse tube refrigerator based cryogenic cooling circuit for NMR cold probes. In: Advances in cryogenic engineering, vol 51A. Springer, Boston, pp 41–48
152. Singh M, Sadana M, Sachdev S, Pratap G (2013) Development of miniature Stirling cryocooler technology for infrared focal plane array. Defence Sci J 63(6):571–580
153. Sreedhar AK, Koteswara Rao KSR (2006) Infrared detectors: materials and technology. Defence Research & Development Organisation, Ministry of Defence, New Delhi
154. Vuillermet M, Billon-Lanfrey D, Reibel Y, Manissadjian A, Mollard L, Baier N, Gravrand O, Brellier D, Destfanis G (2012) Last development of MCT focal plane arrays in France. In: The proceeding of SPIE on electro-optical and infrared systems: technology and applications IX, vol 8541, p 9
155. Stolfi F (1983) Design and fabrication of a long-life Stirling cycle cooler for space application, phases 1 and 2: engineering model, final report. Philips Labs, Inc., Briarcliff Manor
156. Daniels A, Stolfi F, Sherman A, Gasser M (1984) Magnetically suspended Stirling cryogenic space refrigerator: test results. In: Advances in cryogenic engineering, vol 29. Springer, Boston, pp 639–649
157. Wang B, Gan ZH (2013) A critical review of liquid helium temperature high frequency pulse tube cryocoolers for space applications. Prog Aerosp Sci 61:43–70
158. Mai M, Rhlich I, Schreiter A, Zehner S (2011) AIM-space cryocooler programs. Cryocoolers 16:131–141
159. Uhlig K (2002) 3He/4He dilution refrigerator with pulse tube refrigerator precooling. Cryogenics 42(2):73–77
160. Uhlig K (2015) Dry dilution refrigerator with 4He-1 K-loop. Cryogenics 66:6–12
161. Uhlig K (2012) 3 He/4He dilution refrigerator precooled by Gifford-McMahon cooler II. Measurements of the vibrational heat leak. Cryogenics 42:569–575
162. Uhlig K (2008) 3He/4He dilution refrigerator with high cooling capacity and direct pulse tube pre-cooling. Cryogenics 48:511–514
163. Mikheev VA, Noonan PG, Adams AJ, Bateman RW, Foster TJ (2008) A completely self-contained cryogen-free dilution refrigerator, the TritonDR™. Low Temp Phys 34:404–408
164. Prouv T, Luchier N, Duband L (2008) Pocket dilution cooler. Cryocoolers 15:497–503
165. Hata T, Matsumoto T, Obara K, Yano H, Ishikawa O, Handa A et al (2014) Development and comparison of two types of cryogen-free dilution refrigerator. JLTP 175(1/2):471–479
166. Yamanaka Y, Ito T, Umeno T, Suzuki Y, Yoshida S, Kamioka K et al (2009) Development of GM cryocooler separate type liquid-helium-free dilution refrigerator system. J Phys Conf Ser 150:012055
167. Uhlig K (2012) Cryogen-free dilution refrigerator with separate 1K cooling circuit. AIP Conf Proc 1434:1823–1829
168. Oxford Instruments® product catalogue
169. Stirling cryogenics® product catalogue
170. Wang C (2005) Efficient helium recondensing using a 4 K pulse tube cryocooler. Cryogenics 45(12):719–724
171. Wang C (2001) Helium liquefaction with a 4 K pulse tube cryocooler. Cryogenics 41(7):491–496
172. Wang C (2008) Intermediate cooling from pulse tube and regenerator in a 4 K pulse tube cryocooler. Cryogenics 48:154–159
173. Wang C, Abner O (2012) A helium liquefier using three 4 k pulse tube cryocoolers. AIP Conf Proc 1434(1):1640–1646
174. Cryomech® Product catalogue

175. Little WA (1982) Microminiature refrigeration—small is better. Physica 109 & 110B: 2001-2009
176. MMR Technologies Inc. Product catalogue
177. Derking JH, Zalewski DW, Garcia M, Holland HJ, Mudaliar AV, Cao H, Lerou PPPM, ter Brake HJM (2011) Progress in Joule-Thomson microcooling at the University of Twente. Cryocoolers 16:463–471
178. Lerou PPPM (2007) Micromachined Joule-Thomson cryocooler. PhD thesis, 2007, Print: Ipskamp Print Partners, Enschede, ISBN 90-365-2458-x
179. Cao HS, Holland HJ, Vermeer CH, Vanapalli S, Lerou PPPM, Blom M, ter Brake HJM (2013) Characterization of a two-stage 30 K Joule-Thomson microcooler. J Micromech Microeng 23:065022
180. Burger JF, Holland HJ, Seppenwoolde JH, Berenschot E, ter Brake HJM, Gardeniers JGE, Elwenspoek M, Rogalla H (2001) 165 K microcooler operating with a sorption compressor and a micromachined cold stage, Cryocoolers 11. Kluwer Academic/Plenum Publishers, New York, pp 551–560
181. Lerou PPPM, Venhorst GCF, Berends CF, Veenstra TT, Blom M, Burger JF, ter Brake HJM, Rogalla H (2006) Fabrication of a micro cryogenic cold stage using MEMS-technology. J Micromech Microeng 16:1919–1925
182. Lerou PPPM, Veenstra TT, Burger JF, ter Brake HJM, Rogalla H (2005) Optimization of counter flow heat exchanger geometry through minimization of entropy generation. Cryogenics 45:659–669
183. Chao HS, Lerou PPPM, Mudaliar AV, Holland HJ, Derking JH, Zalewski DR, ter Brake HJM (2011) Analysis of multi-stage Joule-Thomson microcoolers. Cryocoolers 16. ICC Press, Boulder
184. Lerou PPPM, ter Brake HJM, Burger JF, Holland HJ, Rogalla H (2007) Characterization of micromachined cryogenic coolers. J Micromech Microeng 19:1956–1960

Chapter 2
Joule Thomson Cryocoolers and Cryoablation

Harrison M. Skye and John M. Pfotenhauer

Abstract Joule Thomson (JT) cryocoolers are finding increased use in the medical application of cryoablation. A high-pressure gas, flowing through the recuperator and orifice of a Joule Thomson cycle produces cooling at the inside tip of a catheter, and an ice ball consequently forms in the surrounding tissue where the catheter tip is located. In one of the most common applications, the catheter tip is positioned at the site of cancerous tissue and when activated, the cooled catheter tip and associated ice ball destroy the cancerous tissue. Upon warming back to normal temperature, the dead cells are then absorbed by the body. The details and advantages of using a cryoablation system based on a two-stage mixed-gas JT cycle are presented. However, a variety of commercial products for cryoablation are available, varying significantly in their technical approach. Multiple product examples presently in use within the USA are described both in terms of their specific JT cycle and working fluid, as well as the associated hardware configurations. High success rates and the associated advantages of this medical procedure such as no-pain and no-scarring combine to suggest a very promising future for cryoablation.

2.1 Overview of Cryosurgery and Cryosurgical Probes

Cryosurgery is a technique for destroying undesirable tissue such as cancers using a freezing process. Treatments include prostate, breast and liver tumor ablation, as well as a variety of dermatological and gynecological procedures. Cryosurgery involves inserting a cryoprobe into the tissue to create the necessary cryogenic temperatures; the cryoprobe tip reaches approximately 150 K and the surgery may last anywhere from a few minutes to an hour [1]. These handheld surgical

H. M. Skye
HVAC&R Equipment Performance Group, Engineering Laboratory, National Institute of Standards and Technology – Gaithersburg, Gaithersburg, MD, USA

J. M. Pfotenhauer (✉)
Department of Mechanical Engineering, University of Wisconsin - Madison, Madison, WI, USA
e-mail: pfot@engr.wisc.edu

M. D. Atrey (ed.), *Cryocoolers*, International Cryogenics Monograph Series,
https://doi.org/10.1007/978-3-030-11307-0_2

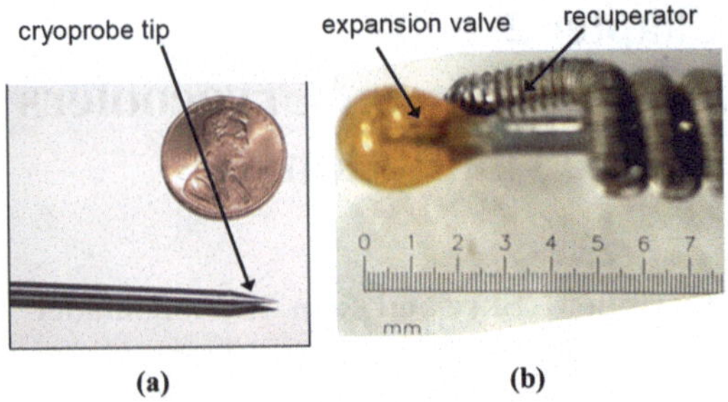

Fig. 2.1 Photos of miniature cryoprobe components including the (**a**) the tip and (**b**) the expansion valve and recuperator of a probe energized by a Joule Thomson cycle

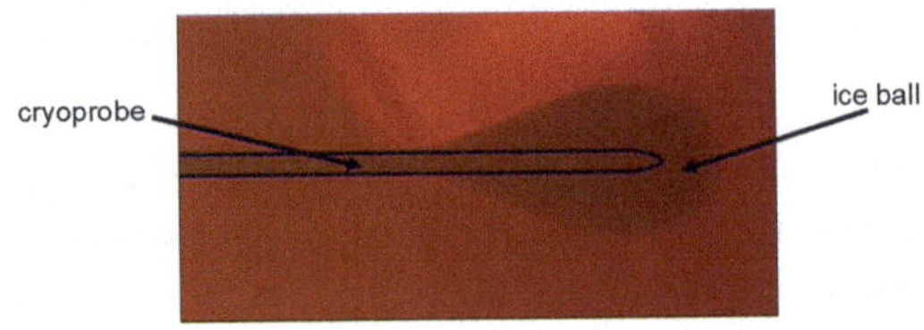

Fig. 2.2 Photograph of an ice ball grown in a gelatin solution using a cryoprobe (Fredrikson [2])

instruments must be compact and ergonomic to facilitate precise placement and to ensure the procedure is minimally invasive. Figure 2.1 shows the components of a particularly small cryoprobe energized by a Joule Thomson (JT) cycle, and demonstrates the level of miniaturization available with this technology.

The cryolesion that is formed has been studied by Fredrikson [2] and is typically on the order of tens of millimeters in diameter. The lethal zone (i.e., the region in which cell death is complete) extends outward into the tissue from the cryo-active portion of the probe approximately to the location where the tissue temperature is about 240 K, although this will vary by ±15 K depending on the details of the surgical procedure and location [1]. The cryolesion is pear-shaped, as shown in Fig. 2.2 (the outline of the probe has been enhanced in the figure to clarify the boundary between the probe and the ice ball). The cryosurgical procedure is inherently less invasive than other treatments as the affected tissue extends beyond the contact point of the instrument. Cryosurgery is therefore an attractive alternative for procedures where surgical resection is not possible because of the proximity of the diseased tissue to large, healthy blood vessels, which may become damaged using a more invasive technique [3].

Cryosurgical treatment of cancers began in the mid-nineteenth century when James Arnott [4] investigated the use of freezing for the treatment of cancer. Freezing tissues using a mixture of ice and various solutes had been previously used as an anesthetic, but Arnott found that freezing was also an effective treatment option for tumors in the breast and uterine cavity [1]. Advances in cryogenics over the next century led to the availability of various cryogens including liquid oxygen and liquid nitrogen as well as solid carbon dioxide (dry ice). However, instrumentation for medical cryogen application was limited during this time and generally capable of freezing to a depth of only a few millimeters [1]. Therefore, the use of cryogenics in medicine was primarily limited to treatment of superficial tissues in the fields of dermatology and gynecology.

Irving Cooper and Arnold Lee [5] invented the first cryosurgical probe that was capable of producing sizable cryolesions deep within the body. Liquid nitrogen (LN2) was pumped through thin concentric tubes; liquid nitrogen entered the probe where it was evaporated by the surgical load at the tip and then nitrogen vapor exited. Liquid nitrogen cryoprobes are still used today; however, the nitrogen vapor is not recovered in the cycle and therefore requires ventilation to avoid an asphyxiation hazard, and the cryogen storage tanks must be periodically refilled which limits the duration of the procedure and adds other logistical complexity. Additionally, the probes and other equipment involved in transporting the liquid nitrogen to the cryoprobe must be vacuum insulated and therefore the system is bulky and difficult to precisely handle; these are undesirable properties for a piece equipment that is meant to be minimally invasive and used in a surgical setting.

The next generation of cryosurgical probes uses a pure gas (e.g., argon) in a Joule Thomson (JT) refrigeration cycle. A high-pressure (often 20 MPa or 3000 psig) gas cylinder is used to provide high-pressure gas to an open-cycle JT system where the low temperature gas in the tip of the cryoprobe creates the cooling effect. The advantage of this system is that the gas entering the cryoprobe is at room temperature and therefore vacuum insulation is not required; these probes are much smaller than their liquid nitrogen counterparts. However, the pressures required by the single component gas in a JT system are too large to be provided by any portable compressor and thus the need for a high-pressure gas bottle. The low-pressure gas leaving the open system is not recovered and therefore represents an asphyxiation hazard; the medical facility must be equipped with an auxiliary ventilation system. The system consumes a large amount of gas since the cooling effect per unit of gas is small; the gas cylinders must be replaced frequently.

JT systems utilizing a mixture of gases, rather than a pure gas, represent a significant advance in cryosurgical probe technology. The pressure required by a mixed gas Joule Thomson (MGJT) system is much lower than for a pure gas JT system. Typically the supply pressure for a MGJT system is 1.5 MPa or 200 psi—an order of magnitude smaller than pure gas systems. Therefore, it is possible to recover the low-pressure mixture leaving the probe and recompress it in a small, portable compressor placed in the operating room. MGJT systems are closed systems that offer the considerable advantage of not using a consumable working fluid; this advantage reduces the hardware, floor space, logistical and ventilation requirements, and expense associated with a procedure. Brodyansky et al. [6] showed that MGJT systems can provide substantially more cooling per unit mass than pure gas JT systems, which leads to a relatively compact and convenient device that is more appropriate for a clinical environment. The thermodynamics underlying the MGJT cycles is discussed in Sect. 2.2 along with associated configurations suitable for use in a cryoprobe.

The current clinical limitations on the use of cryosurgery are primarily related to the cryoprobe technology itself. For treatments that cover large regions deep within the body, current cryoprobe technology requires that multiple probes be inserted and precisely positioned in order to ensure complete cell death. Clearly, a single probe with more power in the same geometric envelope is more desirable as it is less invasive and more easily controlled. The most recent advancement in cryosurgical probe technology addresses this need by improving the underlying thermodynamic cycle. Multi-stage Joule Thomson cycles are used to divide the large temperature range that must be spanned (from room temperature to approximately 150 K) into two smaller temperature stages that can each be addressed using a more compact system. The result is a probe that can provide more refrigeration in the same compact configuration.

2.2 MGJT Cryoprobes and Cycles

Figure 2.3 provides a schematic of a single-stage MGJT cryoprobe configuration and Fig. 2.4a shows the primary components in the single-stage MGJT thermodynamic cycle including numbered thermodynamic states. The compressor and aftercooler deliver high pressure and approximately room temperature gas mixture to the recuperator at state 3. The high-pressure mixture is cooled by the returning

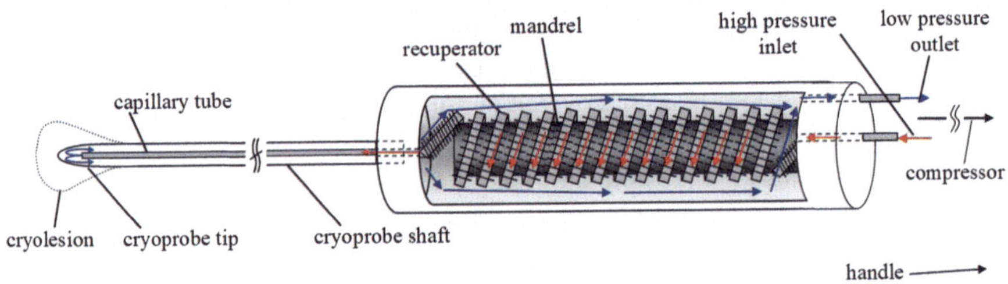

Fig. 2.3 Geometric schematic of a single-stage MGJT cryoprobe

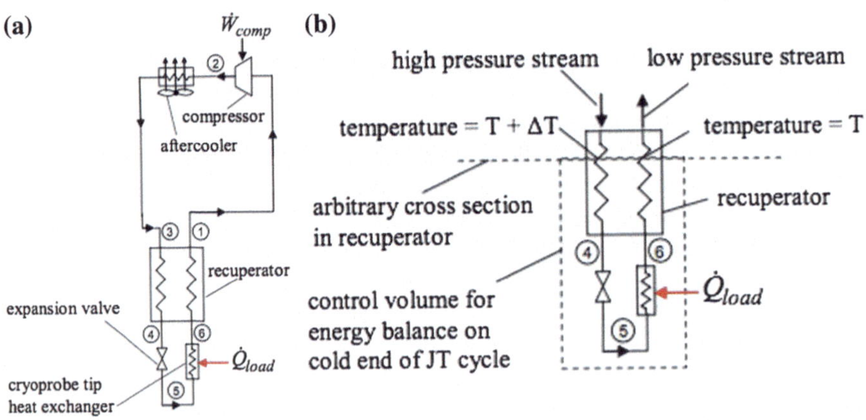

Fig. 2.4 (**a**) Schematic of single-stage MGJT refrigeration cycle. (**b**) Control volume around cold end of JT cycle

low-pressure stream in the recuperator; this heat exchange process enables the cycle to efficiently provide cooling at low temperatures. Isenthalpic expansion across the expansion device (capillary tube, orifice, etc.) reduces the mixture temperature to the lowest temperature in the cycle at state 5. The biological thermal load ($\dot{Q}_{\text{load}}$) is then applied to the flow stream at the cryoprobe tip (represented in Fig. 2.4a as the load heat exchanger); the temperature after the tip (T_6) is typically referred to as the load temperature. The low-pressure mixture then flows through the recuperator and finally returns to the compressor for recovery. The working fluid in the MGJT cycles is typically a Hydrocarbon (HC) or Synthetic Refrigerant (SR) based blend, where the balance gases are noble gases such as nitrogen, krypton, or argon. As discussed in Sect. 2.1, the mixture enters and exits the base of the cryoprobe near room temperature; therefore, the mixture can be transported to and from a remotely located compressor via small and flexible plastic tubing.

The refrigeration capacity of a JT cycle is fundamentally limited by the Joule Thomson effect associated with the working fluid. The capacity can be computed by performing an energy balance on a control volume that encloses the cold end of the cycle; Fig. 2.4b shows a control volume that passes through an arbitrary location in the recuperator and encloses the expansion device and load head exchanger.

The energy balance shows that the refrigeration load is equal to the enthalpy difference between the two streams at any cross section in the heat exchanger:

$$\dot{Q}_{\text{load}} = \dot{m}\left[\text{enthalpy}(P_{\text{low}}, T, \bar{y}) - \text{enthalpy}(P_{\text{high}}, T + \Delta T, \bar{y})\right] \tag{2.1}$$

where $\dot{m}$ is the mass flow rate, P_{high} and P_{low} are the discharge and suction pressures associated with the compressor (neglecting pressure loss in the recuperator), T is the temperature of the low-pressure stream at the location of the control surface, ΔT is the temperature difference between the streams at the cross section, and $\bar{y}$ is a vector of the molar concentrations of each component in the gas mixture.

In the limit that the recuperator conductance is infinitely large (i.e., the recuperator is providing the maximum possible rate of stream-to-stream heat transfer), the temperatures of the fluid streams will coincide (i.e., ΔT in Eq. (2.1) will approach zero) at some location in the recuperator; this location is commonly referred to as the pinch point. The maximum possible enthalpy difference between the two streams, which is equal to the maximum achievable refrigeration load per unit mass flow rate, can therefore be calculated as the minimum value of the isothermal enthalpy difference evaluated over the range of temperature that is spanned by the recuperator:

$$\frac{\dot{Q}_{\text{load,max}}}{\dot{m}} = \min\left(\underbrace{\left[\text{enthalpy}(P_{\text{low}}, T, \bar{y}) - \text{enthalpy}(P_{\text{high}}, T + \Delta T, \bar{y})\right]}_{\Delta h_T}\right) \quad \text{for } T = T_3 \text{ to } T_6 \tag{2.2}$$

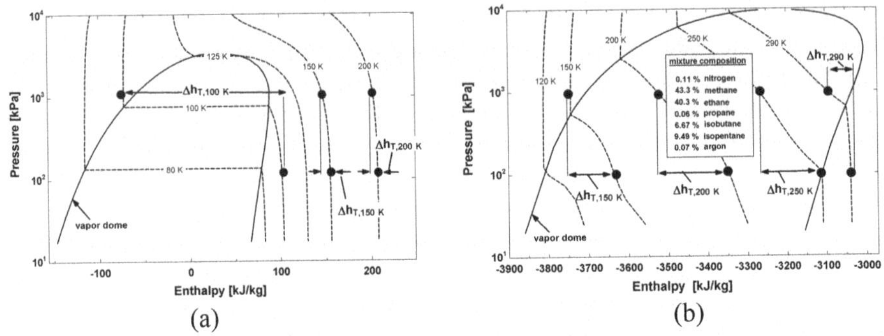

Fig. 2.5 Pressure–enthalpy chart for (**a**) nitrogen and (**b**) an optimized 7-component gas mixture

The isothermal enthalpy difference (Δh_T) is readily evaluated using a pressure–enthalpy (P–h) chart for the working fluid. Figure 2.5a shows a P–h chart for pure nitrogen. Also shown in Fig. 2.5a is Δh_T evaluated for a cycle operating between 1000 and 100 kPa at several different temperatures. Notice that the 100 K isotherm passes through the vapor dome and therefore Δh_T is quite large at this temperature. However at higher temperatures such as 150 and 200 K, nitrogen exhibits behavior that is approaching ideal-gas behavior and therefore Δh_T is very small. This behavior is typical of any working fluid: Δh_T tends to be large only near the vapor dome where real-gas effects govern fluid behavior. The recuperator must nominally span the temperature range from 290 K (warm inlet of recuperator) to 150 K (load temperature – recuperator cold inlet) for a single-stage cryosurgical system. Therefore, the minimum Δh_T will occur at the warm end of the recuperator and will significantly restrict the refrigeration capacity of the cycle. The JT cryoprobe requiring the smallest mass flow rate for a desired cooling power would operate within the vapor dome of the working fluid; however, the recuperator temperature span that is required far exceeds the vapor dome of any single component working fluid.

The vapor dome associated with a zeotropic mixture of gases typically extends over a larger temperature range, corresponding to a temperature that is near the lowest boiling point of the components, to one that is near the highest boiling point of the components. The use of zeotropic gas mixtures therefore significantly extends the temperature range over which Δh_T is large. Figure 2.5b shows a P–h chart for an optimized seven component mixture consisting of nitrogen, methane, ethane, propane, isobutene, isopentane, and argon.

The refrigeration effect (Δh_T) is evaluated using the same pressures, 1000 and 100 kPa, which were used in the nitrogen analysis above. Notice that the values of Δh_T at warmer temperatures are much larger for the mixture because it remains in the vapor dome. The refrigeration effect for pure nitrogen, and the mixture, are shown in Fig. 2.6 as a function of temperature; the refrigeration effect is 50 times longer for the mixture. Cycles that use mixtures can therefore be significantly more powerful and practical for cryosurgery.

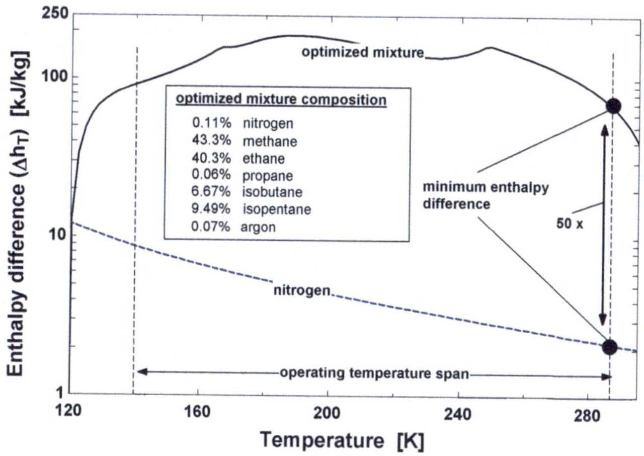

Fig. 2.6 Comparison of isothermal enthalpy difference for nitrogen, and an optimized 7-component mixture

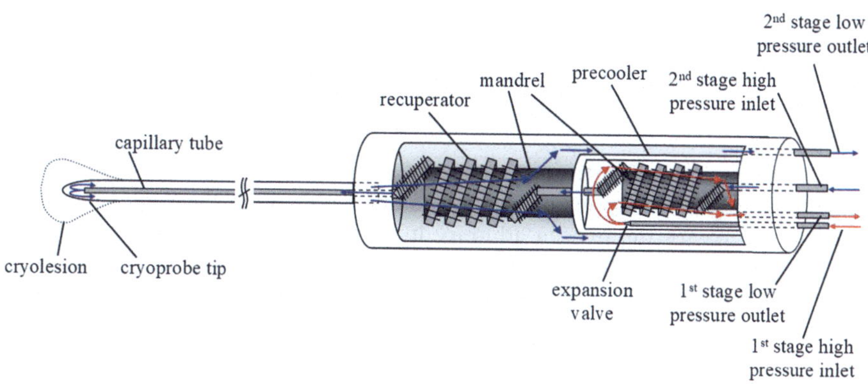

Fig. 2.7 Geometric schematic of a precooled MGJT cryoprobe

2.3 MGJT Cryoprobe with Precooling

The MGJT cycle can be configured to provide even greater refrigeration power using the same physical size of cryoprobe with the addition of a precooling stage (e.g., a 2-stage system). Figure 2.7 shows the physical integration of a precooled MGJT cryoprobe, and Fig. 2.8 provides a cycle schematic of the primary components including numbered thermodynamic states. The "1st stage" is a conventional vapor compression cycle that operates with a single component synthetic refrigerant and precools the high-pressure gas mixture in the 2nd stage JT cycle before it enters the recuperator. The probe configuration is otherwise the same as in the single-stage system where the MGJT cycle provides refrigeration ($\dot{Q}_{load}$) at the tip at the load temperature, T_7.

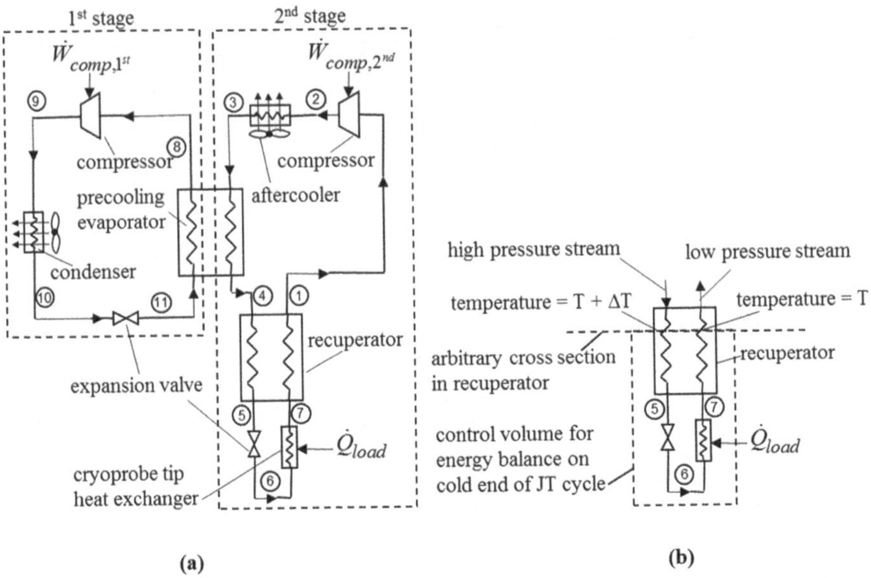

Fig. 2.8 (**a**) Schematic of precooled MGJT cycle. (**b**) Control volume around cold end of the MGJT cycle

The refrigeration effect for this cycle is computed using the same technique as described in Sect. 2.2. An energy balance on the cold end of the 2nd stage of the JT cycle that passes through an arbitrary location in the recuperator is shown in Fig. 2.8b. The energy balance shows that the flow-specific refrigeration load is equal to the enthalpy difference between the two streams at any cross section in the heat exchanger:

$$\dot{Q}_{\text{load}} = \dot{m}_{\text{2nd}}\left[\text{enthalpy}(P_{\text{low,2nd}}, T, \bar{y}_{\text{2nd}}) - \text{enthalpy}(P_{\text{high,2nd}}, T + \Delta T, \bar{y}_{\text{2nd}})\right] \tag{2.3}$$

where $\dot{m}_{\text{2nd}}$ is the mass flow rate in the 2nd stage, $P_{\text{low,2nd}}$ and $P_{\text{high,2nd}}$ are the suction and discharge pressures associated with the 2nd stage compressor (neglecting pressure loss in the recuperator and precooler), T is the temperature of the low-pressure stream, ΔT is the temperature difference between the streams at the cross section, and $\bar{y}_{\text{2nd}}$ is a vector of molar concentrations of each component in the 2nd stage fluid mixture. Again, the maximum achievable refrigeration load per unit mass flow rate is computed as the minimum value of the isothermal enthalpy difference over the range of temperature spanned by the recuperator:

$$\frac{\dot{Q}_{\text{load,max}}}{\dot{m}_{\text{2nd}}} = \min\left(\left[\text{enthalpy}(P_{\text{low,2nd}}, T, \bar{y}_{\text{2nd}}) - \text{enthalpy}(P_{\text{high,2nd}}, T+\Delta T, \bar{y}_{\text{2nd}})\right]\right) \tag{2.4}$$

$$\text{for } T = T_4 \text{ to } T_7$$

The optimized mixture presented in Sect. 2.2 was capable of providing a substantial amount of refrigeration over a large operating temperature span. However, there is a tradeoff between the maximum cooling power that can be provided and the temperature range that must be spanned by the recuperator. For example, consider two different 7-component mixtures that could be used in the JT cycle where the load temperature is 140 K and the high and low pressures are 1000 and 100 kPa. The composition of mixtures A and B has been optimized to produce the maximum JT effect over two different temperature spans but both mixtures have the same constituents: nitrogen, ethane, methane, propane, isobutane, isopentane, and argon. The mole fractions of these constituents are listed in Table 2.1.

Mixture A is the mixture presented in Sect. 2.2 that was optimized for a temperature span of 290–140 K. Mixture B is optimized for a smaller temperature span of 238–140 K that is typical of a JT cycle with some precooling that lowers the recuperator hot inlet temperature to 238 K. Figure 2.9 shows that the maximum

Table 2.1 Mixture operating temperatures and compositions

	Mixture A	Mixture B
Low temp	140 K	140 K
High temp	290 K	238 K
Nitrogen	0.11%	0.0%
Methane	43.3%	50.1%
Ethane	40.3%	39.3%
Propane	0.06%	1.17%
Isobutane	6.67%	9.38%
Isopropane	9.49%	0.01%
Argon	0.07%	0.0%

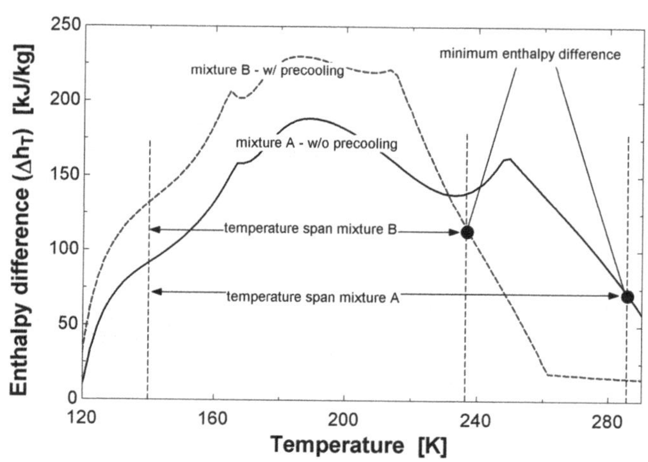

Fig. 2.9 Enthalpy difference of the high (1000 kPa) and low (100 kPa) pressure streams in the recuperator as a function of temperature for two mixtures. The mixtures are optimized to produce the largest cooling effect across two different temperature spans: mixture A 140–290 K, and mixture B 140–238 K

cooling effect (i.e., the minimum value of the isothermal enthalpy change) over the temperature span for mixture A is 73 W/(g/s), whereas the maximum cooling effect for mixture B over its temperature span is 115 W/(g/s).

Therefore, by reducing the temperature range that must be spanned by the recuperator in a mixed-gas JT system, it is possible to select a mixture that achieves a 60% increase in the amount of mass flow-specific refrigeration provided at the tip of the cryoprobe.

A cryoprobe must be compact; that is, a surgically useful cryoprobe will provide a large amount of cooling while still being physically small and therefore surgically ergonomic, minimally invasive, and easy to control. Cryosurgical procedures utilizing a single probe with a high tissue freezing capacity (rather than multiple probes used to simultaneously target a tissue mass) can be carried out more quickly and planned with greater precision. In a single-stage system, the recuperative heat exchanger is rigidly coupled to the shaft of the cryoprobe as shown in Fig. 2.3, and therefore affects the overall cryoprobe size. In the two-stage system, both the recuperative and precooling heat exchangers are coupled to the cryoprobe as shown in Fig. 2.7. Figure 2.10 illustrates the locations of the two heat exchangers in relation to a photo of the precooled MGJT probe donated by American Medical Systems (AMS) to the cryogenics group at UW-Madison. The photo shows that the size of the handheld probe is largely determined by the size of the heat exchangers; therefore, the benefit of precooling must be evaluated based on whether the increase in cooling power outweighs the increase in overall cryoprobe size and the additional complexity associated with the precooling heat exchanger. The size of the heat exchangers is approximately determined by their conductances and so the most appropriate figure of merit for comparing the compactness of different cycles is the ratio of refrigeration load to the total heat exchanger conductance $\left(\frac{\dot{Q}_{\text{load}}}{\text{UA}}\right)$. The conductance of the two-stage system includes the recuperator and precooler, the conductance of the single-stage system only includes the recuperator; the precooled MGJT cycle substantially increases the overall $\left(\frac{\dot{Q}_{\text{load}}}{\text{UA}}\right)$ [7]. The tradeoff is a larger compressor displacement, which, is an acceptable penalty since the compressors are remotely located (provided that the overall probe system remains portable). For example, a 50–60% increase in $\left(\frac{\dot{Q}_{\text{load}}}{\text{UA}}\right)$ can be achieved with a 25% increase in compressor displacement, using a precooling temperature of 235 K.

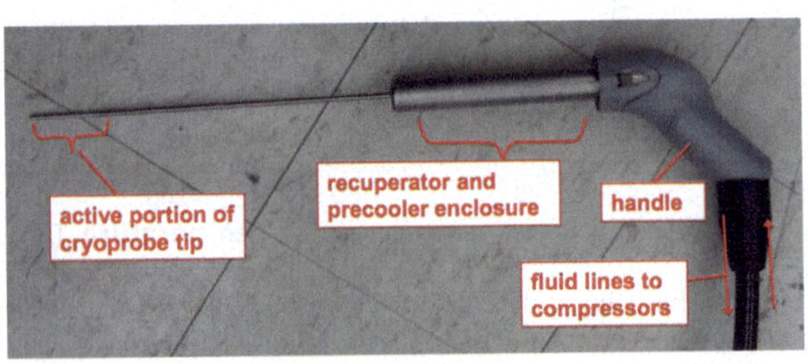

Fig. 2.10 Photograph of the cryoprobe showing the locations of the precooling and recuperative heat exchangers, cryoprobe tip, and the fluid lines which couple the cryoprobe to the compressor cabinet

2.4 Commercial Options for Cryoablation Systems

At the time of writing, there are at least eight US companies providing a catheter-based cryoablation product for treating cancer tumors (prostate, kidney, liver, breast, etc.), atrial fibrillation, or the endometrial lining of the uterus. The assortment of commercial products mirrors the various types of fluid-flow options described in Sects. 2.1–2.3; two of the eight utilize liquid nitrogen pressurized to 0.15 MPa as the working fluid, two utilize argon gas pressurized to slightly higher than 20.0 MPa, two use nitrous oxide (N_2O) pressurized to 5.16 MPa, and two utilize the lower pressure MGJT approach.

Both Sanarus (http://www.sanarus.com/) and IceCure Medical (http://icecure-medical.com/) utilize a supply of liquid nitrogen to produce cooling for cryoablation of breast tumors, and guide the position of the cryoprobe with the help of ultrasound imaging. The probe is inserted through a small (3 mm) incision, after which the slightly pressurized liquid nitrogen is allowed to flow through the probe. Thermal insulation and/or heating along most of the probes length localizes the cooling at the probe tip where the ice ball is formed. The low-pressure nitrogen return flow is exhausted through a connection at the base of the probe and out through the clinic or hospitals scavenging system.

The two companies using room temperature pressurized argon gas to produce cooling at the cryoprobe tip, Galil Medical (http://www.galilmedical.com/) and Endocare, a division of HealthTronics (www.healthtronics.com/) also utilize an alternate flow of pressurized helium gas to re-warm the probe, thereby enabling freeze–thaw cycles. Because the JT inversion temperature for helium gas is well below room temperature (~44 K), the supply of room temperature helium gas at 11.7 MPa flowing through the same cryoprobe tip where the argon had produced cooling, warms on expanding and causes the tip to return to room temperature. The CryoCare CSTM system by HealthTronics and the Visual-ICE system by Galil Medical combine the cryoablation catheters with thermocouples and ultrasound visualization to monitor the cooldown, formation of the ice ball, and warm-up processes. The flexibility of the system allows a selectable cold tip temperature, and the simultaneous use of up to seven probes to increase the extent of the ice ball. Tissue ablation is achieved by repeated freeze and thaw cycles with both freezing and thawing contributing to cell death. Typically, multiple freeze–thaw cycles are used to achieve complete destruction of the target tissue.

Nitrous oxide (N_2O) provides another option as the working gas in the JT cryo-catheter. Medtronic's Arctic Front Cardiac Cryoablation Catheter System (http://www.medtronic.com/us-en/healthcare-professionals/products/cardiac-rhythm/ablation-atrial-fibrillation/arctic-front.html) and Atricure's CryoICE system (https://www.atricure.com/cryo) are both used to treat atrial fibrillation and utilize N_2O as the working fluid. In most medical applications using N_2O, the fluid is stored at room temperature at a pressure of 5.16 MPa in which condition it exists as a saturated liquid. As gas is extracted from the storage cylinder, the mass of liquid decreases, but the pressure remains constant until the last bit of liquid is gone.

Upon expanding to 100 kPa, the gas temperature drops to nearly −80 °C, readily producing a cold zone or ice ball at the catheter tip. Both vendors promote the slow thawing process that occurs after the gas flow is halted as a beneficial method for killing the unwanted cells, noting that a rapid thaw increases the chance of cell survival—a technique that has long been utilized in the treatment of frostbite.

The primary cryoablation product available through Cooper Surgical, Her Option®, (https://www.heroption.com) uses a blend of non-CFC coolants in a MGJT cycle, where the components are non-toxic, non-flammable, and non-corrosive. The Her Option® system is used to ablate the endometrial lining of the uterus in pre-menopausal, post-child bearing women who are experiencing excessive bleeding due to benign causes.

The cooling technology is the same as that developed by American Medical Systems shown in Figs. 2.7, 2.8, and 2.10. The low pressure afforded by the MGJT approach allows the system to operate in the closed cycle described in Sect. 2.3 above, with nominal discharge and suction pressures of 1.5 MPa and 200 kPa, respectively. Because the mixed-gas coolant can be circulated indefinitely, it requires no re-supply of the working gas, a convenient feature that stands in contrast to all the other commercial products. A disposable sheath with a conductive tip is used for each procedure; an electric heater is integrated with the sheath for warm-up cycles. Visualization is accomplished via ultrasound, and the warm-up process typically takes about 2 min. The complete system hardware including the two compressors, controller, and probe is housed in a convenient roll-around unit with outer dimensions of 0.66 m tall by 0.69 m deep by 0.36 m wide. The self-contained package is engineered to provide an exclusively in-office, safe, and effective treatment.

In their Frigitronics® Cryo-Plus™ system, Cooper Surgical also offers an array of cryoprobes that utilize a N_2O approach for cryoablation of cervical or prostate tumors.

2.5 Construction Features

In most cases the cryoprobes used for ablation are fairly simple in construction. Commercial versions avoid any designs reliant on vacuum jackets because of the associated expense to fabricate and guarantee the integrity of such a feature. Single fluid systems such as the liquid nitrogen and nitrous oxide require no recuperative heat exchanger. The simple isenthalpic expansion through the JT orifice at the catheter tip from the supply pressure to near ambient pressure provides a sufficient decrease in temperature to enable formation of an ice ball. Thus, the single-gas version of the cryoprobes are comprised of a concentric set of tubes, with the inner tube providing the flow path for the high-pressure supply, the outer tube providing the flow path for the low-pressure return, and the tip of the tube defining the location for the orifice where the pressure drop and temperature drop occur.

With liquid nitrogen as the working fluid, the concentric tubes are rigid, the outer tube typically being constructed from stainless steel and the inner tube with a thermally low conductivity material such as polyetheretherketone (PEEK) being compatible with the pressure and temperatures involved. Methods to turbulate the return flow enhance the heat exchange between the cold fluid at the probe tip and the surface of the outer tube to form the ice ball (Fig. 2.11). Various methods of warming the return gas, such as an electric heating element, are used to prevent condensation on the exhaust line.

The typical outer diameter of the cryoprobes is a few mm. Examples of rigid and bendable cryoprobes supplied by Atricure are shown in Fig. 2.12a and b. The simple tube-in-tube configuration allows for a section with bellows or malleable metal walls that in turn enable a bendable design; this is useful for applications requiring a particular angle of entry or angle of contact with the patient's organ.

Fully flexible versions of the cryoprobe such as used for treating atrial fibrillation are also simply constructed with tube-in-tube designs using materials such as nylon that can easily bend, and are compatible with the inside lining of blood vessels. In such cases, the cryoprobe diameter is approximately 3 mm, while its length may extend as long as 1.8 m so that it can be directed through blood vessels from an incision in the femoral artery in the groin up to the heart. The pressurized N_2O supply gas expands through an orifice at the distal (far) end of the inner tube producing cooling at the tip of the cryoprobe. Heat exchange between the cooled gas and the tissue surrounding the probe results in the ice ball formation at that location. The cold region is limited to the distal end of the probe by using an insulating sheath or a heating element in the return flow.

For the systems using a room temperature high-pressure gas supply, such as argon, the probe is comprised of a Giauque-Hampson type heat exchanger, that is, a finned tube gas supply line helically coiled around a mandrel, along with a flow-directing outer sheath. The high-pressure supply flows through the inside of the helically wound finned tube. After expanding through an orifice at the tip of the probe, the gas flows back through the space between the coiled tube and the outer wall of the probe toward the base of the probe. The dimensions and insulated length

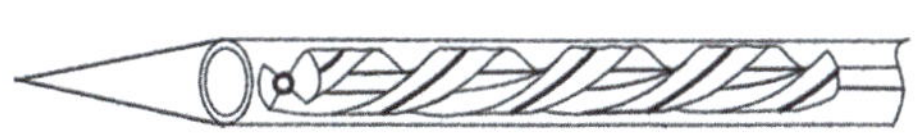

Fig. 2.11 Example of internal configuration to enhance turbulent heat transfer between the return flow and the outer wall [8]

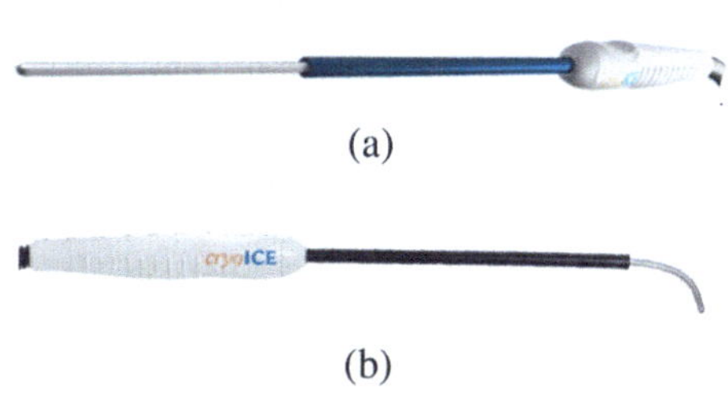

Fig. 2.12 Atricure cryoprobes: (**a**) rigid, (**b**) bendable. Both providing a 10 cm freeze length and utilizing N_2O as the working fluid

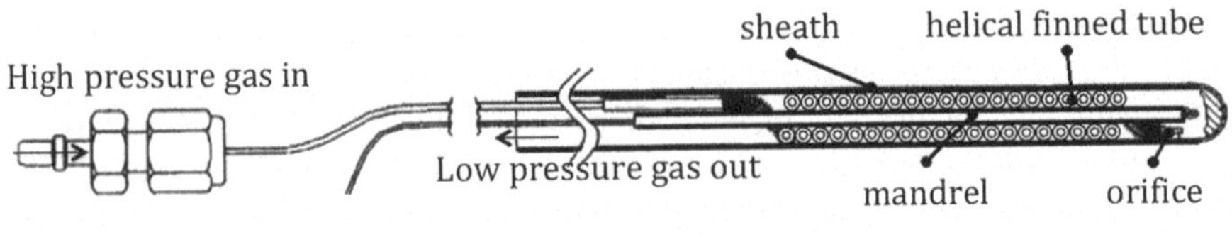

Fig. 2.13 Inner configuration details of a cryoprobe for treating breast tumors

of the outer sheath can be designed to achieve various-shaped ice balls. For example, a sheath with insulation extending over the full length of the coiled heat exchanger, as shown in Fig. 2.13, concentrates the cooling effect at the tip of the probe, thereby forming a spherical ice ball. Such a shape is preferred for treating breast tumors.

On the other hand, a sheath insulation that extends axially only part of the way from the base will result in an elongated cylindrical ice ball such as is used for treating liver tumors. Additional variations on this design including two co-wound parallel gas supply tubes and different locations for the orifice are detailed in Ref. [8]. The pressurized room temperature helium gas used for the warming/thawing process may either flow through the same tube as the cooling gas (argon) or through the second of the two parallel gas supply tubes.

2.6 Precautions

Two precautions typically accompany the use of the cryoablation systems—ensuring the absence of any gas leaks before using the probe, and making sure that the probe is detached from the ice ball before removing it. Both Endocare and Galil include detailed precautionary instructions that each catheter must be pre-tested to ensure no gas leaks are present. With the large supply pressures involved in their systems, greater than 10 MPa, such a safety procedure is crucial and to be expected. One of the patents assigned to Endocare [9] describes a probe including an outer sheath that provides enhanced protection against any gas leaks that would result in a gas embolism. The outer sheath fits over the distal end of the probe and attaches to the probe handle so that any gas leak from the cryoprobe is directed back toward the proximal end. A heat conducting fluid/material such as petroleum jelly fills the very narrow gap between the inner wall of the sheath and outer wall of the probe in order to maintain the probes primary function of creating an ice ball. In one variation of the design, a tube running from the gap between the sheath and probe to a pressure sensor at the proximal (handle) end of the cryoprobe allows the operator to identify any pressure increase from a gas leak and subsequently remove the probe.

Systems that include the two gases, such as argon and helium, to directly regulate the freeze/thaw cycle also prescribe a pre-test of the gas connection. The MGJT system utilized by Cooper Surgical also prescribes a pre-test to verify the freeze cycle performance of the probe, and includes a warning against bending the probe, to avoid any gas leaks.

In the literature provided by each of the companies, explicit warnings are included to not remove the cryoprobe while it is cold. The obvious reason is to avoid tearing any of the tissue adhered to the probe via the ice ball or the cold (<0 °C) probe surface. A variety of approaches are used to re-warm tissue after the ice ball is formed, including: flooding the area with a warm saline solution or other suitable liquid, utilizing the warming effect upon expansion of a flow of pressurized helium gas, or using heater elements to heat the probe or a flow of (non-pressurized) gas through the probe. In many instances operating instructions prescribe multiple freeze–thaw cycles to promote cell death in the target area.

2.7 Future Direction

At the time of this writing, the existing literature regarding the use of cryoablation for treating cancer and other medical problems exudes a very promising future. The procedures report high success rates. Furthermore, the interactive MR Imaging Guided Biopsy (iMR-IGB™) approach being developed by Marvel MedTech is combining MRI imaging with cryoablation to identify and eliminate even pre-cancerous breast tumors. As explained by the first beneficiary of breast cancer cryoablation treatment in the USA, Laura Ross-Paul, and her colleague Ingrid Edstrom of Proactive Breast Wellness (https://www.protectyourbreasts.com), early treatment of breast lesions with cryotherapy before a cancer diagnosis may provide a cancer immunity response by the body. The Sandra and Edward Meyer Cancer Center in New York reports one reason for such an interesting expectation (https://meyercancer.weill.cornell.edu/news/2016-10-19/cryoablation-freezing-away-breast-cancers):

> Besides the no-pain, no-scarring advantages, there's another potential perk with this technique: "The possibility that freezing a cancer and leaving it in the body to be absorbed may stimulate an anti-cancer immune response," explains Dr. Michael Sabel, chief of surgical oncology at the University of Michigan Hospital in Ann Arbor, who also participated in the study. "This has been demonstrated in mice and we are now studying this in patients."
>
> With metastatic breast cancer, in particular, "we have found that if we ablate the original cancer, the metastatic disease goes away in a mouse model," Simmons explains. "What we think is happening is the cryoablation makes the cells burst and they release all that cancer DNA into the system. The body's immune system almost acts as an auto-vaccine against the cancer," which could help reduce the chances of a recurrence as well.

Marvel MedTech further reports (https://www.marvelmedtech.com/the-case-for-imr-igi/) that similar results are being found with humans:

> Cryotherapy—destroying a cancer tumor by freezing—is known to induce a natural anti-tumor immune response [10] that can prevent recurrence and spreading of the disease. Beginning in the early 2000s, research efforts in the U.S. to prove cryotherapy an effective breast cancer treatment have shown promise, but progress here has been slow. Parallel efforts in China have shown greater success, with some 4000 patients treated with cryotherapy and follow-up immunotherapy to boost the anti-tumor immune response. A leading breast cancer research investigator in the U.S., citing recent successes in their understanding of targeted immunotherapy, has predicted that surgical treatment of breast cancer will give way to "Cryoimmunotherapy" as the new breast cancer standard of care. However, at the current pace of progress, it will take at least several more years to achieve this new state of breast healthcare.

A "*natural anti-tumor immune response*" as a result of cryoablation and "cryoimmunotherapy as a new breast cancer standard of care" suggests a very successful future for such procedures. These treatments, along with the ones targeting other types of cancer and atrial fibrillation, combine to suggest a very successful future for cryoablation in general, as well as the Joule Thomson cooling system upon which they are founded.

References

1. Rubinsky B (2000) Cryosurgery. Annu Rev Biomed Eng 2:157–187
2. Fredrikson K (2004) Optimization of cryosurgical probes for cancer treatment. M.S. thesis, University of Wisconsin - Madison, Mechanical Engineering Dept., Madison, WI
3. Zhong-Shan D, Liu J (2010) Numerical study of the effects of large blood vessels on three-dimensional tissue temperature profiles during cryosurgery. Numer Heat Transfer, Part A: Appl 49(1):47–67
4. Arnott J (1851) On the treatment of cancer by the regulated application of an anaesthetic temperature. Churchill, London
5. Cooper IS, Lee AS (1961) Cryostatic congelation: a system for producing a limited, controlled region of cooling or freezing of biologic tissues. J Nerv Ment Dis 133:259
6. Brodyansky VM, Gresin AK, Gromov EM, Yagoden VM, Nicolsky VA, Alpheev VN (1971) The use of mixtures as the working gas in throttle Joule Thomson cryogen refrigerators. In: Proceedings of the 13th international congress of refrigeration. Progress in refrigeration science and technology, Washington, DC, vol 1, p 43
7. Skye H, Nellis G, Klein S (2009) Modeling and optimization of a cascaded mixed gas Joule-Thompson cryoprobe system. ASHRAE transactions, vol 115 Part 2, pp 966–983, 2009. ASHRAE transactions - Papers presented at the 2009 annual conference in Louisville, Kentucky of the American Society of Heating, Refrigerating and Air-Conditioning Engineers, Inc. ISSN: 00012505

8. Mikus PW, Kelly GL, Brady RK (1998) Cryoprobe. US Patent number 5,800,487
9. Mikus PW, Eum JJ (2002) Cryoprobe assembly with detachable sheath. US Patent Application number 09/978,653
10. Mehta A, Oklu R, Sheth RA (2016) Thermal ablative therapies and immune checkpoint modulation: can locoregional approaches effect a systemic response? Gastroenterol Res Pract 2016. Article ID 9251375

Chapter 3
Mixed Refrigerant Joule–Thomson Cryocooler for Cooling High-Temperature Superconductor Cables

Sangkwon Jeong, Jisung Lee, and Cheonkyu Lee

Abstract A cryogenic Joule–Thomson (JT) refrigerator has several advantages for large industrial applications, in the aspect of easy cooling power adjustability and high reliability due to no moving parts at low temperature. A mixed refrigerant Joule–Thomson (MRJT) refrigerator technology, which has usually exhibited good performance at refrigeration temperature above 80 K, can be useful for cooling high-temperature superconducting (HTS) system if it is modified to extend the cooling temperature up to 70 K. The demand of highly reliable, highly efficient, large cooling capacity and low-cost refrigeration for cooling HTS cables has spurred investigation of advancing an MRJT cryogenic refrigerator. A precooling stage is an essential part of an efficient MRJT refrigerator at the refrigeration temperature of 70 K, and a refrigeration cycle including a precooling cycle can be designed and optimized to achieve reasonably high COP. The neon–nitrogen MRJT refrigerator is specifically needed for cooling subcooled liquid nitrogen of the HTS cable. The MR for the first stage (low-temperature cycle) of the cascade MRJT refrigerator and the single MRJT refrigerator is composed of nitrogen (N_2), argon (Ar), tetrafluoromethane (CF_4, R14) and octafluoropropane (C_3F_8, R218). R410A is selected as the refrigerant for the second stage (high-temperature cycle) of the cascade MRJT refrigerator, of which the cooling temperature is approximately

S. Jeong (✉)
Division of Mechanical Engineering, Cryogenic Engineering Laboratory, School of Mechanical, Aerospace and Systems Engineering, Korea Advanced Institute of Science and Technology, Daejeon, Republic of Korea
e-mail: skjeong@kaist.ac.kr

J. Lee
Launcher Propulsion Control Team, KSLV-II R&D Head Office, Korea Aerospace Research Institute, Daejeon, Republic of Korea
e-mail: jisung26@kari.re.kr

C. Lee
Thermal & Fluid System R&BD Group, Research Institute of Sustainable Manufacturing System, Korea Institute of Industrial Technology, Cheonan, Republic of Korea
e-mail: cklee@kitech.re.kr

M. D. Atrey (ed.), *Cryocoolers*, International Cryogenics Monograph Series,
https://doi.org/10.1007/978-3-030-11307-0_3

240 K. With its versatility of constituents and scale, the nonflammable MRJT technology is competitive in commercial sectors with the already developed Stirling and Brayton refrigerators. The technical advancement of the MRJT cryogenic refrigerator is emphasized particularly for the cooling purpose of large-scale HTS application. The optimized MRJT refrigerator in this chapter has demonstrated its potential as a cooling system for HTS cables with improved efficiency at 70 K.

3.1 Introduction

Joule–Thomson (J-T) effect is one of the most frequently utilized methods to decrease temperature in a refrigeration system including cryocooler. It is commonly realized with a simple throttling device such as an expansion valve, a nozzle or a capillary tube. These devices do not have an intended mechanical interaction to produce any work while the expansion process happens too rapidly to exchange heat with the environment. The isenthalpic process is often presumed for such a simple Joule–Thomson expansion, although it is an irreversible process. The Joule–Thomson coefficient is a representative parameter to indicate the quantitative temperature change upon pressure change and is defined as follows:

$$\mu_{\mathrm{JT}} = \left(\frac{\partial T}{\partial P}\right)_h \tag{3.1}$$

Using the thermodynamic relation, it is further expressed in terms of other thermodynamic properties like internal energy of the working fluid. The next derived equation shows that the Joule–Thomson coefficient is zero if the expanded fluid is an ideal gas, because the internal energy of ideal gas is a function of temperature only. Further, it also obeys the ideal gas equation of state, $Pv = RT$.

$$\begin{aligned} \mu_{\mathrm{JT}} &= \left(\frac{\partial T}{\partial p}\right)_h \\ &= -\left(\frac{\partial T}{\partial h_p}\right)\left(\frac{\partial h}{\partial p_T}\right) \\ &= -\frac{1}{C_p}\left[\left(\frac{\partial u}{\partial p}\right)_T + \left(\frac{\partial (pv)}{\partial p}\right)_T\right] \end{aligned} \tag{3.2}$$

Examining the above Joule–Thomson coefficient asserts the fact the significant cooling effect can be only obtained by this simple expansion technique when the expanded fluid or refrigerant is near its saturated condition. We call this as a real gas "J-T cooling effect." The typical temperature–entropy diagram of cryogenic fluid is represented in Fig. 3.1a. The initial cooldown process of a simple Joule–Thomson cryocooler from room temperature using a pure cryogenic fluid is very slow or it may require enormous pressure ratio to produce reasonable cooling effect. This fact is schematically illustrated by Fig. 3.1b by indicating a transient expansion process. The initial expansion of high-pressure cryogenic fluid near room temperature does

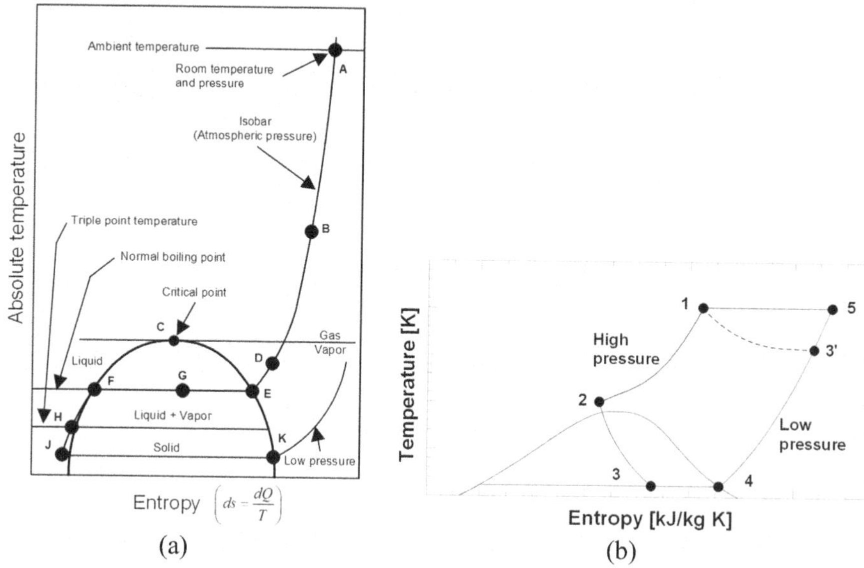

Fig. 3.1 (**a**) Temperature entropy diagram of typical cryogenic fluid and (**b**) ideal J-T cycle of pure refrigerant with recuperator (3′ is the low temperature obtained after initial or intermediate J-T expansion during transient operation while 1–2–3–4–5 are the thermodynamic states when the refrigeration cycle reaches steady state)

not produce sufficient cooling effect because it usually occurs far from its saturation region. Its J-T coefficient is almost zero like an ideal gas, which results in just slight temperature decrease unless extraordinarily high pressure is utilized as shown in Fig. 3.1b. The temperature of the state 3′ is close to that of 5. Repetitive J-T expansions may ultimately lower the expansion temperature so that the steady state cooling temperature can be realized with reasonable cooling capacity that is usually accompanied by an evaporation process. As pointed out precisely by Barron [1], however, simple Linde–Hampson systems cannot even attain low temperatures at all with helium, hydrogen, and neon that have maximum inversion temperatures below room temperature. In this case, a precooling process of the compressed fluid in a cascade manner is indispensable to reach cryogenic temperature.

Figure 3.2a is a schematic diagram of Joule–Thomson cryocooler with a counter-flow recuperative heat exchanger (or recuperator). In general, it is composed of a compressor, an aftercooler, a heat exchanger, an expansion valve, and an evaporator or a low-temperature cooling-load heat exchanger. This particular cryogenic system is considered for cooling HTS cable. In the case of ideal recuperation by the heat exchanger, the potential cooling effect of the low-pressure stream after J-T expansion is to be fully utilized for precooling the high-pressure stream and the specific steady-state cooling capacity becomes the minimum isothermal enthalpy difference of the refrigerant between the cryogenic J-T stage and room temperature. Figure 3.2b shows the schematic J-T cascade refrigeration system where a precooling stage is indispensable for helium, hydrogen, or neon J-T cryocooler. Simple

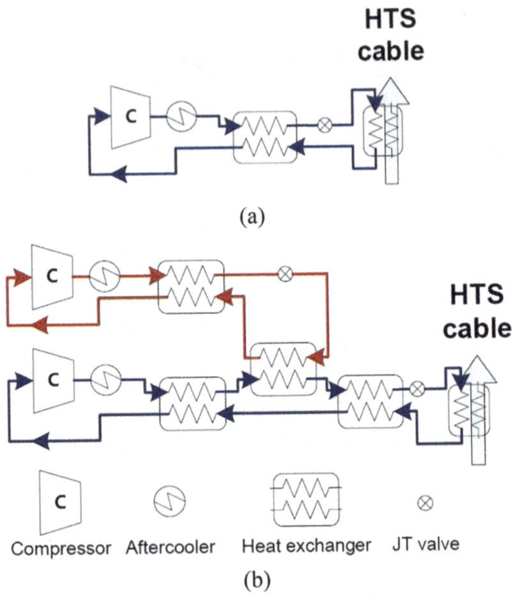

Fig. 3.2 Cycle configuration of (**a**) simple J-T system and (**b**) cascade system

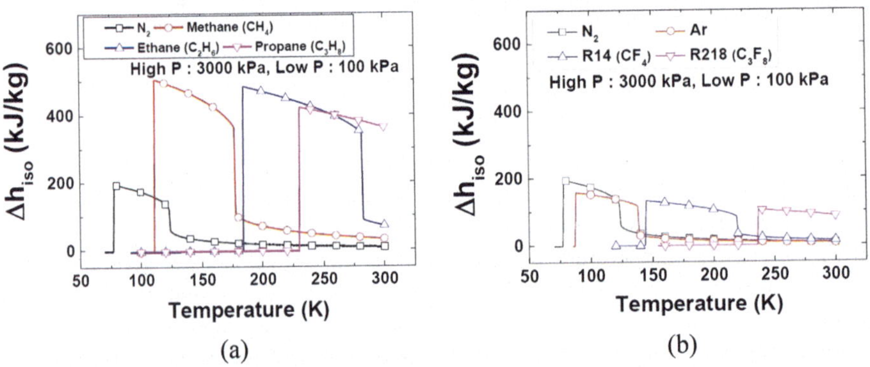

Fig. 3.3 Isothermal enthalpy difference of cryogenic fluid; (**a**) flammable refrigerants with nitrogen, (**b**) nonflammable refrigerants with nitrogen

addition of precooler at appropriate temperature range is actually beneficial for increasing the overall thermodynamic efficiency of cryogenic refrigeration system. As shown in Fig. 3.3, each cryogenic fluid possesses a significant value of pressure-variant enthalpy difference within its specific temperature range only, i.e., in the vicinity of its saturation condition. This means that the selection of refrigerant for J-T cryocooler is very important and has to be carefully done according to its target cooling temperature. For example, propane (C_3H_8) is a thermodynamically excellent refrigerant around room temperature, but not so effective below 220 K. Methane (CH_4) is not so efficient for creating measurable cooling effect around room temperature, but very powerful refrigerant around 120 K. Nitrogen (N_2)

is obviously a poor refrigerant to reduce temperature near room temperature because it behaves almost like an ideal gas, but it is absolutely necessary for producing cooling effect near 100 K or below. The following conjecture, therefore, is justifiable. Precooling of the high-pressure stream before the temperature range of large isothermal enthalpy difference may greatly enhance the performance of J-T cryocooler which has to use a pure component refrigerant. If the target cooling temperature is sufficiently low and the large cooling capacity is demanded, a separate precooling refrigeration cycle, therefore, is necessary as shown in Fig. 3.2b. In the two-stage cascade system as an example, each J-T refrigeration cycle is completely separated so that the different refrigerants are selectively used in two cycles. The main advantage of cascade cycle is its higher thermodynamic efficiency particularly when the temperature span between the heat source temperature and the heat sink temperature is large [2].

On the other hand, the system complexity increases, and thus additional caution is required for the refrigerant leak problem. However, more complicated cascade systems with an extension of several precooling stages were practically used for the early natural gas liquefaction process [3] to increase the liquefaction efficiency by choosing the most efficient and proper refrigerant for each stage. In fact, liquefaction process of natural gas was very well optimized with so-called mixed refrigerant (MR) which has similar components of natural gas. There are numerous LNG (liquefied natural gas) plants that have been well developed in the form of highly efficient cryogenic refrigeration system in terms of thermodynamic sense. C3MR is the most famous MR system which is not only a generic MR cryogenic refrigerator but also a cascade system with the precooling process of propane (C_3H_8).

The great potential advantage of MRJT system is that the target cooling temperature can be lowered flexibly by adding the low boiling temperature component, and the large cooling effect from the high boiling temperature components may be still available. Radebaugh succinctly pointed out the usefulness of MRJT refrigerator by comparing its COP with those of pure N2-JT systems as shown in Table 3.1 [4]. The operating temperature range is assumed from 240 to 75 K, and the J-T expansion pressure is set as 100 kPa. The minimal compression work with an isothermal compression at 300 K is used to estimate the required compression work. For MRJT system, the cooling temperature may not be constant during the heat absorption in the evaporator by temperature glide characteristic of mixture. Nonetheless, however, the significantly increased Carnot efficiency reveals the benefit of the MRJT system. Another advantage of MRJT system is the low compression ratio. For this example,

Table 3.1 Efficiency of N_2 and MRJT system ($T_H = 240$ K, $T_L = 75$ K, $P_L = 100$ kPa)

Refrigerant	P_H (kPa)	$(\Delta h_T)_{min}$ (kJ/mol)	COP_{ideal}	Carnot efficiency (%)
N_2	5000	0.468	0.048	14.4
N_2	2500	0.232	0.0289	8.7
N_2	1600	0.146	0.0211	6.3
MR	1600	1.576	0.228	68.4

the system COP of MRJT system is higher than any other N2-JT systems even though the pressure ratio is as small as 16. Most of MRJT cryocoolers show high efficiency at compressor discharge pressure approximately 2500 kPa which can be easily obtained from commercial oil-lubricated refrigeration compressors. Thus, the cost of MRJT cryocooler operating with commercially available compressors can be reduced to $3000 [5].

Despite relatively difficult maintenance and handling of refrigerant in case of leakage, MRJT system is attractive due to its high thermodynamic efficiency. As previously discussed, the most notable application of MRJT system is found in natural gas liquefaction plants. Propane precooled MRJT system (C3-MR) or dual MRJT system (DMR) is currently the most efficient liquefaction cycle in LNG (liquefied natural gas) industry [6]. Small-scale MRJT cryocoolers are also extensively used. Over 1000 units of MRJT cryocoolers are adopted and in operation for gamma-ray detectors at about 90 K [5]. A tiny expansion part of the MRJT cryocooler smaller than a coin size was developed for the heart arrhythmia cryogenic catheter [7, 8]. By the use of optimized mixed refrigerant, it was possible to realize the cryoprobe to only 3 mm in diameter. Recently, a micro-scale MRJT cryocooler was successfully developed for cooling small electronic devices at about 150 K. Multipurpose commercialized coolers such as Polycold compact coolers [9] and MMR Kleemenko coolers [10] are also operating as MRJT cycle, and they are widely used in laboratories for various cooling temperatures.

The aim of this chapter is to explore the possibility of MRJT cryocooler for cooling superconducting devices such as high-temperature superconductor (HTS) cable. The target cooling temperature is typically lower than 77 K, because the HTS cable has to be cooled by subcooled liquid with sufficient temperature margin to avoid cavitation and require enough dielectric strength of coolant. Although numerous MRJT refrigerators have been developed for efficient cooling systems above 120 K, the efficiency noticeably decreases below 100 K. Thus, in order to overcome such low thermodynamic efficiency of MRJT cryocooler below 77 K, a cascade concept must be adopted [11]. As shown in Fig. 3.4, the main cooling cycle utilizes a neon–nitrogen MR for creating 64 K while the precooling cycle may be constituted with another MRJT system. All of them are not accompanied with a phase separator for the sake of simple operation. It may be optional if one needs to design a precooling cycle with a single- or double-stage MRJT system. According to our simulation results, the cascade precooling cycle of MRJT as depicted in Fig. 3.4 substantially enables the whole cooling system to achieve high thermodynamic efficiency. Table 3.2 compares several candidate cooling methods for HTS cable at 70 K. Although Stirling cryocooler and Brayton refrigerator are considered mature technologies that are readily available for cooling at 70 K or below, the well-designed MRJT cryocooler can give rise to a competitive performance value. They are also safe and favorable for scalability because the cooling power demand for 1 km or longer superconducting cable system will typically require an order of 10 kW level. The electric utility side should favor a reliable system that has few moving parts at cryogenic temperature, and the MRJT system should be able to scale up with less hardware overhead burden.

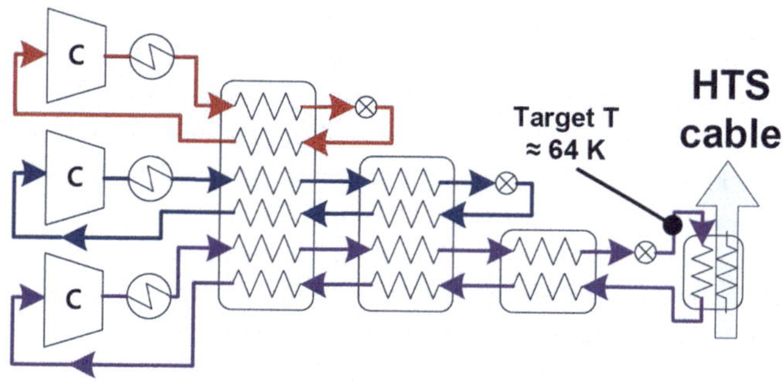

Fig. 3.4 Possible arrangement of cascade MRJT cryocooling system for HTS cable

Table 3.2 Efficiency comparison for HTS cable cooling system ($T_H = 300$ K, $T_L = 70$ K)

Refrigeration cycle	COP	Carnot efficiency %
MRJT without precooling	0.004	1.3
Precooled MRJT [12]	0.053–0.064	17.4–21
Stirling refrigerator [13]	0.065	21.3
Brayton refrigerator [14, 15]	0.05–0.08	16.5–26.3

3.2 MRJT Refrigerator for Cooling Below 77 K

As described in the previous section, most MRJT cryocooling systems have been developed in conjunction with LNG industry or biomaterial storage which typically produce cooling effects above liquid nitrogen temperature. This section specifically examines some research efforts to extend the cooling temperature range of MRJT below 77 K since that is the cooling temperature requirement of an HTS cable. A suction pressure of commercial compressors is usually maintained higher than 100 kPa. Accordingly, the refrigeration temperature below 77 K cannot be attained with pure nitrogen because its saturation temperature is 77 K at 100 kPa. Mixing non-condensable gases such as neon or helium, therefore, can be the solution to reach below 77 K in this suction pressure constraint. However, these fluids decrease the overall efficiency of refrigeration cycle because their J-T coefficients are negative at ambient temperature [1]. For these reasons, the usual MRJT refrigerators could exhibit good performance only above 77 K [16]. Prior researches on the MRJT refrigerator, however, have been conducted to explore the refrigeration temperature below 77 K. In 1973, nearly 63 K of refrigeration temperature was achieved by Alfeev et al. [17] using neon–nitrogen–methane–ethane–propane (25/25/15/15/20% by vol.) mixture, but the efficiency of refrigerator was not reported. Since then, Khatri et al. [18] conducted experiments with various mixtures for the purpose of cooling HTS devices. As a result, 5.8% of Carnot efficiency was achieved at 91 K, and the minimum temperature or no-load temperature was 68 K without any cooling load. In 1997, no-load temperature of 69 K was achieved by Luo et al. [19] with neon and hydrocarbon mixture again, and the Carnot efficiency was ranged between 5% and 9% at 77 K. These values of Carnot efficiency are significantly improved compared to pure nitrogen J-T refrigerator of which Carnot efficiency is only between 0.5% and 1.5%. In 1999, no-load temperature of 64 K was achieved by Arkhipov et al. [20] using neon, nitrogen, and hydrocarbon mixture, and the Carnot efficiency was between 2.5% and 5% at 80–85 K range. In the other research effort, no load temperature of 51 K was achieved by Luo et al. [21] using a phase separator that was a different cycle from those mentioned above. Although neon–nitrogen–oxygen–argon (40/15/35/10% by mole) mixture was used as the mixed refrigerant, all high-boiling-point components except neon were returned to the high-temperature section by the phase separator which was installed at the precooling stage, so that only neon was expanded at the last expansion valve. This experiment was conducted with an open cycle and hence the information about the system efficiency was not reported.

In summary, it was successful to reach a refrigeration temperature below 77 K by mixing neon in the refrigerant. However, the common problem of low efficiency of MRJT refrigerator at this temperature level has been remained as a challenging problem.

3.3 NF (Nonflammable) MRJT Refrigerator for Cooling Below 77 K

There are two additional factors to take into account in developing an MRJT cryocooler for HTS device: flammability and the resulting freezing problem. Although hydrocarbon mixtures with nitrogen or neon were capable of serving as good MRs in various industries, the electrical utility is very conservative and concerned about a potential hazard issue such as fire, explosion, or flammability. An application for cooling HTS cable dictates the strict requirement of safety or avoiding any potential hazard. We, therefore, have to specifically utilize nonflammable (NF) MR for the J-T refrigeration system of HTS cable. Besides its poor thermodynamic efficiency, using nonflammable refrigerant as the working fluid of a cryocooler usually presents an inherent thermodynamic issue; too high triple point temperature as shown in Table 3.3. The liquid temperature spans of potential NF refrigerants in a J-T cycle are considerably narrower than those of hydrocarbon refrigerants because they are constituted between the critical and triple point temperatures. This means that the freezing temperature of useful NF fluid as MR may be problematic to reach cryogenic temperature. Nevertheless, due to the solubility effect, some mixed refrigerant can show remarkably low freezing temperature which is lower than each component's individual triple point temperature. Thus, the solid-liquid equilibria

Table 3.3 Triple point temperatures of various nonflammable pure refrigerants

Refrigerants	Triple point temperature (K)
R14 (CF_4)	89.5
R23 (CHF_3)	118.0
R218 (C_3F_8)	125.4
R245fa ($CHF_2CH_2CF_3$)	171.1
R125 CHF_2CF_3	172.5
Nitrous oxide (N_2O)	182.3

for NF MR are primary research contents for developing an MRJT cryocooler, especially for HTS device. As illustrated in Fig. 3.4, the main J-T refrigerator with neon–nitrogen MR does not cause a freezing problem as long as it stays above the triple point of nitrogen which is 63.2 K [22]. Since the critical temperature of neon is only 44.5 K, we cannot expect and also do not need any solubility effect to lower the freezing temperature in this case. The main focus of mitigating a potential freezing problem is the MR of the precooling cycle which typically needs to cool the neon–nitrogen MR of the main cycle around 100 K in the cascade arrangement of MRJT system for HTS cable. The MR of argon (Ar), tetrafluoromethane (CF_4, R14), trifluoromethane (CHF_3, R23), and octafluoropropane (C_3F_8, R218) is considered as one of the most suitable combination for a nonflammable MRJT refrigerator with the aim of cooling below 120 K such as natural gas liquefaction.

For multicomponent MRs, the ternary mixture set of Ar, R14, and R218 are explored extensively in conjunction with the development effort of low-temperature MRJT system. The exclusion of R23 is due to its high freezing temperature in the mixture, which is discussed in the previous research [23]. The previous research explains that R23 is the main component of freezing problem in the temperature above 110 K. Furthermore, the freezing point of the selected ternary MR (Ar, R14, and R218) has been measured extensively [24]. The freezing point depression was clearly observed due to its mixing characteristic in the whole MR cases, which was absolutely compared to that of pure R218. This depression behavior strongly appears with a certain MR which has R14-rich composition. If the ternary MR contains R14 fraction higher than 0.5, no freezing phenomenon occurs even at 77 K! Also, the molar ratio between R14 and R218 is a key parameter to drop the freezing temperature. If this molar ratio is higher than 2 between R14 and R218, the freezing state of MR does not appear even at liquid nitrogen environment (77 K). In other words, the liquid state of ternary MR of Ar, R14, and R218 can maintain the temperature below 77 K. Therefore, these selected MRs with extraordinary low freezing temperature characteristic can favorably operate the NF MRJT refrigerator as the precooling cycle without any freezing trouble of the working fluid itself.

3.4 Configuration of the Designed NF MRJT Refrigerator for Cooling HTS Cable

Figure 3.5 illustrates the practical cooling scheme of HTS cable with MRJT system where simple two-stream counterflow heat exchangers are utilized. In order to cope with the cooling task of HTS cable below 77 K which is the boiling temperature of liquid nitrogen at 1 bar, the exact composition of the components of MR should be determined first according to the demand of HTS cable system.

In general, the refrigeration temperature below 77 K cannot be achieved with pure nitrogen at a compressor suction pressure of 1 bar. This problem can be solved by lowering the partial pressure of nitrogen by adding a fluid with a critical point lower than nitrogen such as neon, hydrogen, and helium. Hydrogen has been excluded due to its combustibility. As the J-T coefficient of neon is much larger than that of helium, a neon–nitrogen mixture has been selected as the MR of the main cycle [12]. Nitrogen, methane, ethane, and propane have been initially selected as the components of MR in the precooling cycle because they are the most common MR components in the LNG industry. However, to fulfill the nonflammable requirement of refrigerant as addressed in the previous section, the flammable refrigerants are excluded. The quaternary MR of nitrogen (N_2), argon (Ar), tetrafluoromethane (CF_4, R14), and octafluoropropane (C_3F_8, R218) is primarily considered instead [25]. The precooling refrigeration cycle has been designed as shown in Fig. 3.6 in order to minimize the complexity of heat exchanger design. To increase the overall thermodynamic efficiency of the precooling MR cycle, an additional R410A precooling stage is added in Fig. 3.6, which takes major latent heat load of the high pressure MR stream. It is evident that this extra precooling component is optional and also other common refrigerant can be used instead of R410A for cooling approximately at 230 K.

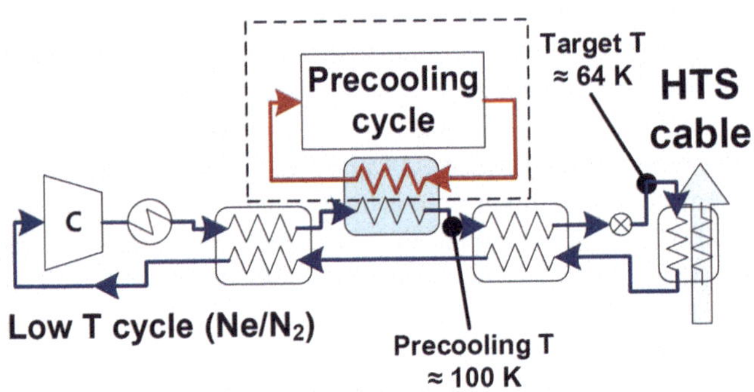

Fig. 3.5 Schematic diagram of the MRJT refrigeration system for HTS cable cooling by using two-stream heat exchangers only

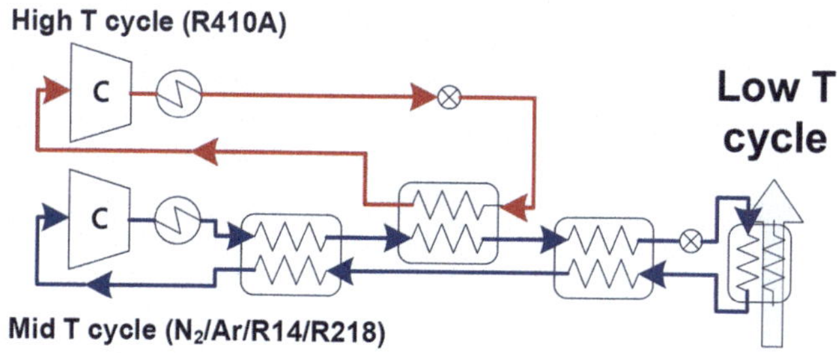

Fig. 3.6 Schematic diagram of the cascade precooling MRJT refrigeration system with an R410A precooling stage

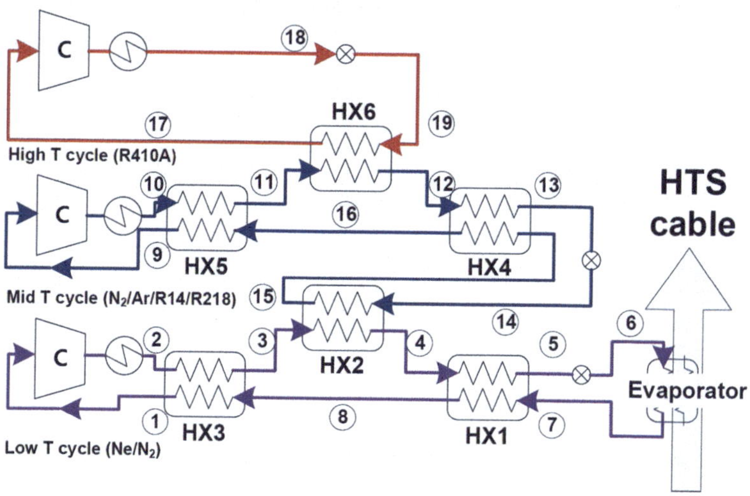

Fig. 3.7 Cycle configuration of MRJT refrigerator with LRS mode operation

3.5 Design Analysis Results of the Practical NF MRJT Cryogenic Refrigerator for HTS Cable

Figure 3.7 shows the schematic diagram of the practical HTS cable cooling system with cascade MRJT system.

The main cycle which utilizes neon–nitrogen MR (42/58% mole) has been designed to satisfy the cooling requirement of liquid nitrogen at 70 K, and its detailed temperature and pressure conditions are listed in Table 3.4.

In the case of precooling cycle with the optimum composition of quaternary MR, i.e., 20% nitrogen, 30% argon, 30% R14, and 20% R218 mixture, the detailed design parameters are also listed in Table 3.5.

The maximum COP of the precooling cycle is obtained as 0.216 as shown in Fig. 3.8 which indicates the optimized composition calculation results.

Table 3.4 Temperature, pressure, and vapor fraction of the neon–nitrogen MRJT cycle shown in Fig. 3.7 ($Ne/N_2 = 0.42/0.58$)

Stream	Temperature (K)	Pressure (kPa)	Vapor fraction (kg/kg)
1	297.0	100	1.000
2	300.0	2300	1.000
3	116.8	2300	1.000
4	102.2	2300	0.765
5	75.0	2300	0.342
6	64.1	100	0.471
7	70.0	100	0.662
8	99.0	100	1.000

Table 3.5 Temperature, pressure, and vapor fraction of the precooling MRJT cycle shown in Fig. 3.7 ($N_2/Ar/R14/R218 = 0.2/0.3/0.3/0.2$ for the stream 9–16, and R410A for the stream 17–19)

Stream	Temperature (K)	Pressure (kPa)	Vapor fraction (kg/kg)
9	297.0	220	1
10	300.0	1400	1
11	253.9	1400	0.974
12	247.8	1400	0.921
13	108.4	1400	0
14	97.6	220	0.033
15	105.2	220	0.267
16	240	220	1
17	250.9	287	1
18	300	1750	0
19	244.8	290	0.346

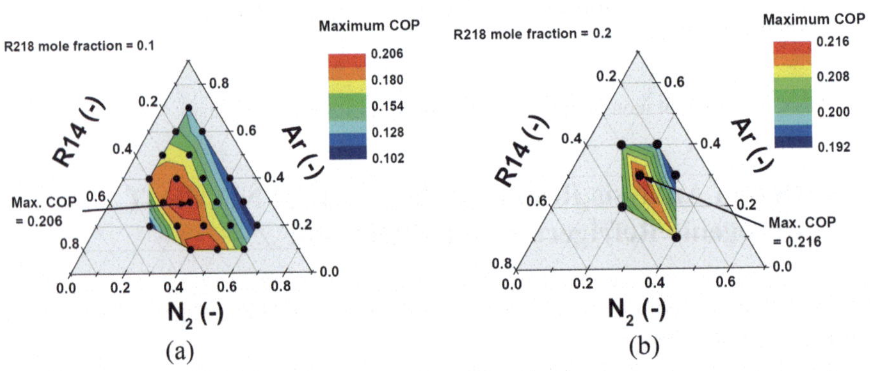

Fig. 3.8 The maximum COP comparisons of the quaternary cascade MRJT refrigerator for various MR (**a**) mole fraction of R218 = 0.1 and (**b**) mole fraction of R218 = 0.2

The overall COP of the MRJT refrigerator is greatly influenced by the COP of precooling cycle as shown in Fig. 3.9 and compared with the commercialized Stirling and Neon Brayton cryocoolers [12].

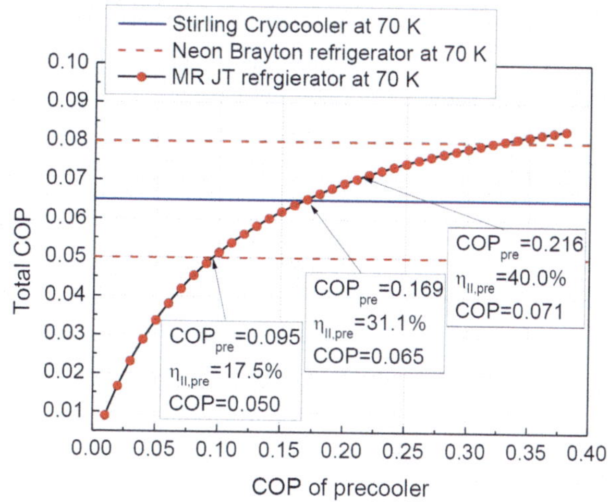

Fig. 3.9 Overall COP of MRJT refrigerator according to the COP of precooling cycle

It is clear that the MRJT refrigerator is a feasible system for cooling HTS cable, when it is properly assisted by an MR precooling cycle. It has comparable thermodynamic efficiency with other cryogenic refrigerators while it is free of moving parts at low temperature. The design analysis did not consider any pressure drop effect in heat exchangers. Therefore, like any other cryocooler design or fabrication, the heat exchanger is surely one of the most important components that determine the overall system efficiency. All the analyses in this chapter assume the minimum temperature difference of heat exchanger as 3 K or larger. The design concept is in principle to generate less entropy due to finite temperature difference in heat exchangers by considering temperature profiles associated with heat capacity rate of high- and low-pressure streams. The other factor which may significantly influence the refrigerator system performance is the inherent isentropic efficiency of the compressor. Our analyses are based on the assumption that the readily available compressors can be utilized for the specifically designed MR composition with isentropic efficiency of 80%.

3.6 Reliability of Cascade System

The cycle configuration of multistage MRJT cycle is more complex than the single-stage refrigeration cycle. The main concern is the reliability of the increased number of compressors. The reliability of the compressor can be expressed in failure rate from the reliability database [26]. Another caution is the possibility of the refrigerant leak problem in the increased number of compressors and the heat exchangers. In the database, the leak frequency is classified by the hole diameter [27]. The leak frequency from the historical data [28, 29] and the compressor failure rate are summarized in Table 3.6.

Table 3.6 Compressor failure rate and leak frequency data

Equipment	Failure rate per 1,000,000 h	Active repair time (h)
Electric motor drive	87.3	47.5
Compressor	256.4	25.7
Equipment	Hole diameter (mm)	Leak frequency per year
Compressor	0–2	4.63×10^{-3}
	2–10	6.72×10^{-3}
	10–50	5.76×10^{-3}
Heat exchanger	0–2	4.95×10^{-3}
	2–10	1.15×10^{-3}
	10–50	3.52×10^{-3}

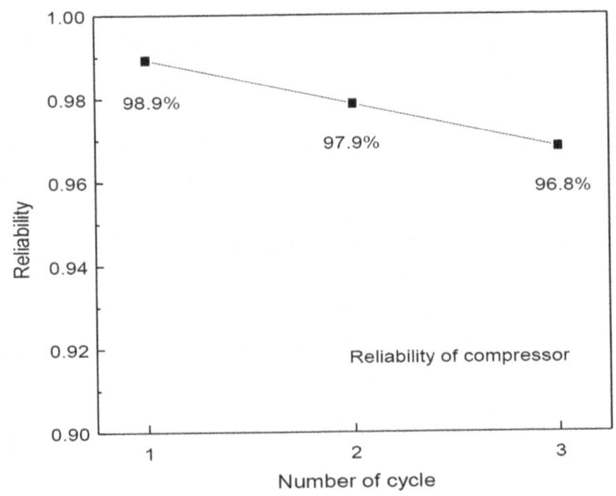

Fig. 3.10 Reliability of compressor according to the number of cycle

The reliability of compressor is calculated by subtracting the active repair time in a year when the compressor failure occurs. The variation of the compressor reliability according to the number of cycle is presented in Fig. 3.10.

The reliability of compressor decreases with the increase of cycle number; however, the reliability is maintained as high as 96.8% even in the three-stage refrigeration cycle. The refrigerant leak frequency in the compressor and the heat exchanger is also presented in Fig. 3.11 with the number of cycle.

In this calculation, the refrigeration cycle is assumed as operating with multi-stream heat exchanger. The leak frequency in the compressor is higher than that in the heat exchanger, thus the highest probability of leak is 0.036 at the compressor in the three-stage refrigeration cycle. By the conversion, the refrigerant leak accident may occur once in 28 years even with the three-stage refrigeration cycle. In conclusion, the reliability degradation by the addition of precooling stages is not significant in the aspect of the compressor failure and the leak problem. It is evident that this is one of the reasons that ConocoPhillips cascade LNG liquefaction process is still working in Kenai [30].

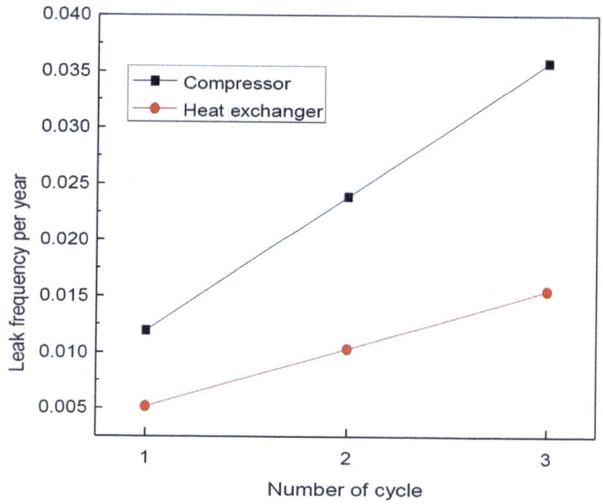

Fig. 3.11 Refrigerant leak frequency in the compressor and the heat exchanger according to the number of cycle

3.7 Concluding Remarks

The demand of highly reliable, highly efficient, large cooling capacity, and low-cost refrigeration for cooling HTS cables has spurred investigation of advancing an MRJT cryogenic refrigerator. A precooling stage is an essential part for an efficient MRJT refrigerator at the refrigeration temperature of 70 K, and a refrigeration cycle including a precooling cycle can be designed and optimized to achieve reasonably high COP. With its versatility of constituents and scale, the NF MRJT technology is competitive in commercial sectors with the already developed Stirling and Brayton refrigerators.

The freezing problem of low-temperature nonflammable MR is also addressed and extensively discussed with freezing temperature depression issue. This chapter quantitatively analyzed the COP of MRJT refrigerator for the potential use of cooling HTS cable, and the achievable COP of the MRJT refrigerators is summarized. The developed MRJT refrigerator is featured to have a high COP when it is composed of two-stage precooling cycle. Although the complexity of the MRJT cycle is increased by the addition of a precooling stage, the characteristics of high reliability and low cost are still conserved due to the absence of mechanically moving expansion parts. The technical advancement of the MRJT cryogenic refrigerator is emphasized particularly for the cooling purpose of large-scale HTS application. The optimized MRJT refrigerator in this chapter has demonstrated its potential as a cooling system for HTS cables with improved efficiency at 70 K.

References

1. Barron RF (1985) Cryogenic systems. Oxford University Press, New York
2. Jeong S, Smith J Jr (1994) Optimum temperature staging of cryogenic refrigeration system. Cryogenics 34(11):929–933
3. Conoco Phillips Liquefied Natural Gas (2016) http://lnglicensing.conocophillips.com/what-we-do/lng-technology-licensing/Pages/optimized-cascade-process-advantages.aspx
4. Radebaugh R (2010) Foundation of cryocoolers short course. In: International cryocooler conference, vol 16
5. Radebaugh R (2009) Cryocoolers: the state of the art and recent developments. J Phys Condens Matter 21(16):164219
6. Frg W, Bach W, Stockmann R, Heiersted R, Paurola P, Fredheim A (1999) A new LNG baseload process and production of the main heat exchanger. Linde Rep Sci Technol 61:3–11
7. Marquardt E, Radebaugh R, Dobak J (1998) A cryogenic catheter for treating heart arrhythmia. In: Advances in cryogenic engineering. Springer, New York, pp 903–910
8. Dobak J, Xiaoyu Y, Ghaerzadeh K (1998) A novel closed loop cryosurgical device. In: Advances in cryogenic engineering, vol 43. Springer, New York, pp 897–902
9. Polycold systems. http://www.brooks.com/support/technical-support/documentation/cryochillers-cryocoolers/~/media/Files/Support/Documentation/Vacuum/WaterpumpChillers/Manual%20Cryocoolers/82510100.pdf
10. MMR technologies (2016) http://www.mmr-tech.com/mmr_overview.php
11. Alexeev A, Haberstroh C, Quack H (2002) Mixed gas J-T cryocooler with pre-cooling stage. In: Cryocoolers 10. Springer, New York, pp 475–479
12. Lee J, Hwang G, Jeong S, Park BJ, Han YH (2011) Design of high efficiency mixed refrigerant Joule-Thomson refrigerator for cooling HTS cable. Cryogenics 51(7):408–414
13. Stirling cryogenics BV (2016) http://www.stirlingcryogenics.com
14. Yoshida S, Hirai H, Takaike A, Hirokawa M, Aizawa Y, Kamioka Y, Okamoto H, Hayashi H, Shiohara Y (2010) New design of neon refrigerator for HTS power machines. In: Advances in cryogenic engineering, vol 55. Springer, New York, pp 1131–1138
15. Hirai H, Hirokawa M, Yoshida S, Sano T, Ozaki S (2014) Development of a turbine-compressor for 10 kW class neon turbo-Brayton refrigerator. In: Advances in cryogenic engineering. Springer, New York, pp 1236–1241
16. Radebaugh R (2004) Refrigeration for superconductors. Proc IEEE 92(10):1719–1734
17. Alfeev VN, Brodyanski V, Yagodin V, Nikolsky V, Ivantsov A (1973) Refrigerant for a cryogenic throttling unit. UK patent 1(336):892
18. Khatri A (1996) A throttle cycle refrigerator operating below 77 K. In: Advances in cryogenic engineering. Springer, New York, pp 1291–1296
19. Luo E, Liang J, Zhou Y, Yakuba V, Lobko M (1997) Experimental investigation of an efficient closed-cycle mixed-refrigerant J-T cooler. In: Cryocoolers 9. Springer, New York, pp 529–535
20. Arkhipov V, Yakuba V, Lobko M, Yevdokimova O, Stears H (2002) Multicomponent gas mixtures for J-T cryocoolers. In: Cryocoolers 10. Springer, New York, pp, pp 487–495
21. Luo E, Gong M, Zhou Y, Zhang L (2000) Experimental investigation of a mixed-refrigerant J-T cryocooler operating from 30 to 60K. In: Advances in cryogenic engineering, vol 45(A). Springer, New York, pp 315–322
22. Lee J, Oh H, Jeong S (2014b) Investigation of neon-nitrogen mixed refrigerant Joule-Thomson cryocooler operating below 70K with precooling at 100K. Cryogenics 61:55–62
23. Lee C, Lee J, Yoo J, Jeong S, Jung J (2014a) Investigation of visualized solid-liquid phase equilibria for pure and mixed refrigerant. In: Miller SD, Ross RG Jr (eds) International cryocooler conference 18, Syracuse, New York, USA, 2014a. Boulder, Colorado, pp 397–406
24. Lee C, Yoo J, Lee J, Jeong S (2016) Visualization of the solid-liquid equilibria for non-flammable mixed refrigerants. Cryogenics 75:26–34

25. Lee C, Jin L, Park C, Jeong S (2017) Design of non-flammable mixed refrigerant Joule-Thomson refrigerator for precooling stage of high temperature superconducting power cable. Cryogenics 81:14–23
26. OREDA (2002) Offshore reliability (OREDA) handbook. Det Norske Veritas (DNV), Hvik
27. Nam K, Chang D, Chang K, Rhee T, Lee I-B (2011) Methodology of life cycle cost with risk expenditure for offshore process at conceptual design stage. Energy 36(3):1554–1563
28. Health GB, Executive S (1997) Offshore hydrocarbon releases statistics 1997: 1/10/92 to 31/3/97 inclusive. Great Britain, Health and Safety Executive
29. Spouge J (1999) A guide to quantitative risk assessment for offshore installations. CMPT
30. Andress D, Watkins R (2004) Beauty of simplicity: Phillips optimized cascade LNG liquefaction process. In: Advances in cryogenic engineering, vol 49. Springer, New York, pp 91–98

Chapter 4
Cryocooler Applications at Neutron Scattering Facilities

Oleg Kirichek

Abstract A global shortage of helium gas can seriously jeopardise the scientific programmes of neutron scattering laboratories due to the use of cryogenic sample environment in the majority of the neutron scattering experiments. Recently developed cryogen-free technology allows a significant reduction or even a complete elimination of liquid helium consumption. Here we review the impact of the cryogen-free revolution on cryogenic equipment used at large neutron facilities, such as cryostats, dilution refrigerators, superconducting magnets and other cryogenic systems.

4.1 Introduction

In the last decade, neutron scattering facilities have experienced a booming popularity. Two major neutron sources, SNS in the United States and J-PARC in Japan, have already started operating, and other major facilities such as ESS (European Spallation Source) and CSNS (China Spallation Neutron Source) are at advanced stages of construction. Leading players in the neutron scattering field such as ILL in France, ISIS in the UK or FRM II in Germany have significantly upgraded their capabilities and continue to carry out further facility development programmes. This, rapidly growing interest in neutron scattering, can be explained by astonishing new opportunities offered to users of neutron scattering facilities, due to considerable progress in a number of areas such as neutron optics, neutron detection, large and complex dataset analysis, neutron scattering instrumentation and sample environment.

Today, two thirds of all neutron scattering experiments are performed under cryogenic conditions [1, 2]. There are fundamental reasons for that. Firstly, the thermal motion of atoms is reduced at low temperature, significantly improving

O. Kirichek (✉)
ISIS Neutron and Muon Source, Science and Technology Facilities Council,
Rutherford Appleton Laboratory, Didcot, UK
e-mail: oleg.kirichek@stfc.ac.uks

M. D. Atrey (ed.), *Cryocoolers*, International Cryogenics Monograph Series,
https://doi.org/10.1007/978-3-030-11307-0_4

the precision of structural characterisation of materials. Secondly, a cryogenic temperature range allows the study of low temperature phase transitions.

A cryogenic environment is also required to provide the high magnetic field sample environment generated by superconducting magnets [3], or for operating of neutron moderators (where the energies of neutrons get reduced to usable level) [4] and some components of neutron beamline instruments, such as cryogenic beryllium filters [5] or cryopumps [6].

Traditionally at neutron scattering facilities, the cryogenic environment is used to be provided by liquid cryogen-based systems; however, extensive use of this kind of equipment requires significant resources, including the considerable cost of the required cryogens and a number of logistic issues, and poses health and safety problems.

In the middle of the 2000s, neutron scattering facilities have been exposed to another serious threat—global helium shortage. In the period from 2008 until 2012, some of the commercial helium traders occasionally suspended the liquid helium supply to research organisations. For neutron facilities, this problem is of special importance, because during the last two decades, they are permanently in the top-ten list of liquid helium users in the scientific world, somewhere after large accelerators, fusion reactors and MRI and NMR laboratories. Large-scale neutron facilities like ILL, SNS or ISIS used to consume tens of thousands litres of liquid helium annually. This situation posed one of the major risks on sustainability of a neutron facility's operation. Another negative consequence of helium supply shortage is the decade-long rise in helium prices. During the period from 2005 to 2015, the cost of liquid helium in the UK has quadrupled. That increase created a significant additional burden on facilities budgets. There is also very serious environmental aspect of this problem. According to William Halperin: "Helium gas is a natural resource which is not replaceable and when released it will rise through, and escape from our atmosphere and be gone forever" [7].

Luckily the new ground-breaking cryogen-free approach emerged at the beginning of the twenty-first century, offering an effective solution. So, what is the essence of this approach?

In the case of a conventional cryostat, the cooling power required to maintain the cryogenic conditions is provided by evaporating liquid cryogens: helium and nitrogen. In the case of cryogen-free systems, the cooling power is provided by a closed cycle refrigerator (CCR), also known as a cryocooler. There are two options available in order to reduce or, in the ultimate case, eliminate the usage of cryogens. The first option is comprised of the so-called dry systems, which do not contain liquid cryogens at all, and are built around a CCR, utilising the cooling power produced by the cold head [8–11]. The second option is based on the idea of recondensing the evaporating helium back to the cryostat by a CCR [12].

The breakthrough became possible due to a new generation of CCRs developed in late 1990s and early 2000s. Initially cryogen-free systems were based on conventional Gifford-McMahon (GM) cryocoolers [13]. However, despite obvious advantages, such as operational simplicity, the use of these systems was limited by the high level of mechanical vibrations produced by GM cryocoolers, and the necessity of expensive and regular service inspections. A new generation of CCRs based on the pulse tube refrigerator (PTR) technology offers an effective solution to both problems [14, 15]. A unique feature of the PTR is the absence of cold moving parts. This considerably reduces the noise and vibrations generated by CCR, a critical issue for a number of scientific instruments. Furthermore, the PTR offers lower maintenance costs and less disruption, as expensive high-precision seals are not required, and the cold head can be operated without a service inspection.

Out of all the large-scale scientific facilities, cryogen-free technology produced most of its impact on neutron facilities. This is because the approach based on CCRs is only efficient for small cryogenic devises, like cryostats, but not for bigger cryogenic systems, such as large superconducting magnets or cavities of accelerators. For the latter, industrial scale liquefiers are much more effective.

In the last couple of years, the helium shortage situation has noticeably eased and helium prices have stabilised; however, the CCR technology continues to expand, mostly due to its operational simplicity and reliability. At ISIS facility, between 60 and 70% of all user experiments that require cryogenic sample environment use CCR-based systems. Currently there are 57 operational cold-heads at ISIS facility, and this number continues to grow.

In this chapter, we briefly review the impact of CCR technology on the major cryogenic systems in a neutron facility experimental hall: cryostats, dilution refrigerators, superconducting magnets and neutron beamline instrument components. Naturally all cryogenic systems described in this chapter fall into two major groups: mobile cryogenic devises and stationary systems. Each of them utilises different advantages of CCR technology. Mobile cryogenic devises benefit from the compactness of CCR-based systems and a relatively quick turnover, while stationary systems exploit low maintenance and low resource requirements, which are characteristic for this technology.

4.2 Cryogen-Free Cryostats

A cryostat is a device that allows to maintain the cryogenic environment of a sample, for as long as required. In modern cryogenics, there is a great variety of different types of cryostats, each with its own advantages and disadvantages relative to others. Since the late 1970s, the Orange cryostat designed at the ILL [16] has been the workhorse of all neutron scattering facilities. This is a top-loading, liquid helium bath cryostat, with a liquid nitrogen-cooled infrared radiation shield, and a base temperature of ~1.5 K. However, at the beginning of the twenty-first century, new cryostats based on cryogen-free technology have emerged. Even though all these cryostats, developed at different facilities, have the same basic principles of operation, the concept has generated many individual designs [8–11, 17]. Photos of an example of a top-loader system (based on PTR) developed by ISIS, and manufactured by AS Scientific, are presented in Fig. 4.1 [10, 11]. The system consists of (1) the outer vacuum vessel; (2) the top-loading insert with an internal

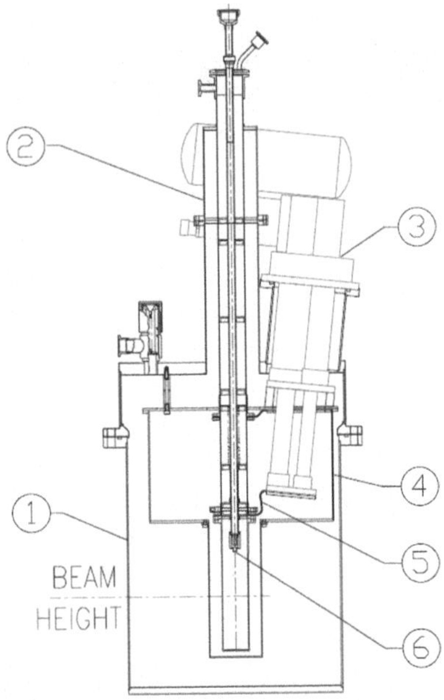

Fig. 4.1 Diagram of the pulse tube top-loading system

diameter of 50, 75 or 100 mm; (3) the CCR; (4) the infrared radiation shield attached to the first stage of CCR and (5) the thermal link between the second stage and the insert base flange. A sample is loaded into the insert on the end of a sample stick. The infrared radiation shield is made of highly conductive copper and is covered with high emissivity aluminium foil. The neutron beam access to the sample is provided through the thin aluminium or vanadium foil windows in the cryostat tails. In order to enable laser alignment of samples, sapphire windows might be fitted on the reflectometer instruments.

For all designs of CCR-based top-loading cryostat, those based on PTR systematically reach lower base temperatures (3.0–3.5 K) than those based on GM (4.3–6 K).

The system cooling time of under 3 h, for PTR based systems, is also shorter than the approximate 4 h for the GM-based systems. The vibration level measured in the PTR cryostat is an order of magnitude less than that in a similar system based on a standard GM cryocooler [10].

For a rough estimation of the CCR-based top-loading cryostat efficiency, we can compare the cost of liquid helium required to run a standard Orange cryostat at ~60/day (based on liquid helium price in 2016) with the cost of the electric energy consumed by a CCR ~10/day [17]. This difference can only rise in the foreseeable future.

Currently ISIS has 11 PTR-based and 8 GM-based top-loading cryostats and 21 Orange cryostats. Most neutron facilities, including ISIS, also have a number of bottom-loading CCR cryostats which can be used as a cryo-furnace with temperature range 2.8–700 K [9]. The capacity of the fleet of cryogen-free cryostats allows running more than half of all the low-temperature sample environment experiments at ISIS.

However, experiments which need temperatures lower than 2 K still use Orange or similar cryostats. At ILL, a third-stage Joule–Thomson refrigerator added to the GM 10 K cryocooler reduced the temperature down to 1.8 K [1]. Addition of a similar Joule–Thomson stage to the CCR-based top-loading cryostat, for reaching 1.5 K, was later suggested in [11]. This idea has been realised in a prototype of the 1.5 K cryogen-free, 50 mm diameter, top-loading cryostat based on a PTR, which has been developed and successfully tested at ISIS [11–19]. The sample temperature range of this system is 1.4–300 K in the continuous flow regime. A sustainable high cooling power of 0.23 W at 1.9 K and temperature stability of 0.1 K across the temperature range 1.4–200 K has been achieved. This top-loading cryogen-free cryostat may also be used with ultra-low temperature inserts. Thanks to the high cooling power of the cryostat's VTI heat exchanger, the cryostat temperature range can be extended to the millikelvin area by inserting a standard dilution refrigerator insert in the VTI. In our tests using the Kelvinox dilution insert, a base temperature of 55 mK has been achieved in the continues circulation regime [19]. All the parameters of the new system are very close to the ones of conventional cryogen-based cryostats, such as the Orange cryostat. Furthermore, the new system does not require cryogenic liquid top-ups, which are challenging from the logistical, economical, operational and safety points of view.

High interest in cryogen-free cryostats in the neutron scattering community has stimulated industrial companies to develop commercial products for neutron scattering sample environments based on the technology. Currently Oxford Instruments, Ice Oxford, AS Scientific, Cryogenics, Janis and other companies are offering variety of cryostats based on cryocoolers.

4.3 Powerful Cryogen-Free Dilution Refrigerators Used in Neutron Scattering Experiments

A significant number of experiments at advanced neutron scattering facilities require sample temperatures below 1 K. Such sample environments are usually provided by dilution and/or 3He evaporation refrigerators. Many dilution refrigerators available at neutron facilities are based on the design incorporating sintered silver heat exchangers [20–22]. For many years, this used to be the standard of conventional dilution refrigerators. Usually this kind of dilution refrigerators are built in a liquid helium bath cryostat, with all the consequences typical for liquid cryogen-based cryogenics.

The development of the cryogen-free dilution refrigerator started at the very beginning of the twenty-first century, alongside the development of cryogen-free technology in general [23–25]. In most case, the design of modern, powerful, cryogen-free dilution refrigerators is based on the prototype developed by Kurt Uhlig [26], where the PTR is used to precool the dilution unit. These powerful dilution refrigerators are usually used in experiments with heavy and large sample

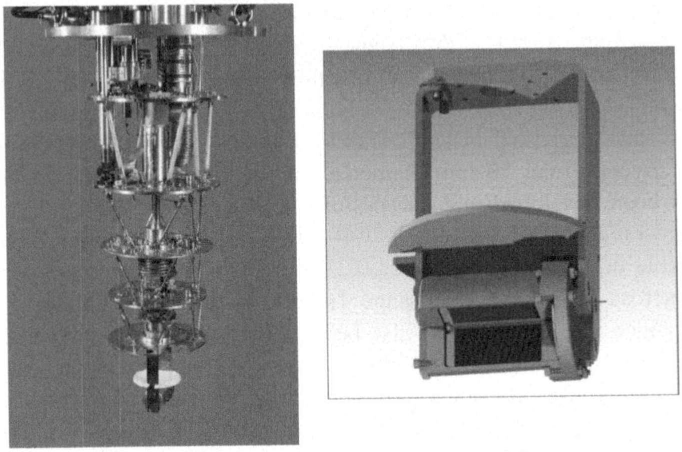

(a) (b)

Fig. 4.2 (**a**) Powerful cryogen-free dilution refrigerator E-18; (**b**) 110 cm^3 solid 4He sample cell (also shown attached to the mixing chamber in (**a**))

cells, like those used for high-pressure measurements. The photo of a similar powerful cryogen-free dilution refrigerator (E-18) developed by Oxford Instruments in collaboration with ISIS [2] is presented in Fig. 4.2a. The fridge is capable of cooling large (diameter 200 mm; height 250 mm) and heavy (up to 20 kg) samples and provides access for the neutron beam through a set of thin aluminium alloy windows. The base temperature of the refrigerator is 15 mK and the cooling power is 370 W at 100 mK. In experiments where E-18 has been used to cool down 110 cc solid and liquid 4He sample cells (Fig. 4.2b; also attached to the mixing chamber in Fig. 4.2a), the base temperatures of 35 mK for the solid sample and 50 mK for the liquid sample (up to 25 bar) have been held for weeks [27–29]. To our knowledge, the paper published in [27] became the first publication where a powerful cryogen-free dilution refrigerator has been used in a neutron scattering experiment.

Today dilution refrigerators of this type are commercially available from a number of companies such as Leiden Cryogenics [30], Air Liquide [31], Oxford Instruments [32], LTLab [33], Cryoconcept, Janis and BlueFors.

4.4 Cryogen-Free Advanced Superconducting Magnets for Neutron Scattering Experiments

Neutron scattering is an invaluable tool for solid state magnetism. In many experiments, a sample needs to be exposed to a high magnetic field, low temperature and a neutron beam simultaneously [34]. Nowadays, these conditions can be satisfied only in superconducting magnets. First generation of superconducting magnets for neutron scattering has been developed in late 1980s, and by the middle of 1990s, the design had settled down and magnets became a part of the standard sample environment (SE) kit [35]. Most of them are split-pair magnets made of NbTi and Nb3Sn filamentary superconducting wires. The magnet is usually built in a conventional liquid helium bath cryostat, with a thermal shield cooled by liquid nitrogen. A set of windows made of neutron transparent materials provide a neutron beam access to the sample.

However, in the last decade, the rapid development in superconducting magnet technology and cryogenics has made it possible to build the new generation of advanced magnets for neutron scattering SE [3, 36–40]. The new magnets are usually designed to satisfy specific requirements of a particular advanced neutron scattering instrument, such as magnet aperture/opening tailored to instrument detectors coverage or compatibility with special collimating systems. This allows to create an optimal combination of instrument and magnet, which can achieve much faster experiments with much higher resolution.

Here we discuss two superconducting magnets, cryogenics of those is based on helium recondensing by a PTR cryocooler: the 14 T magnet for neutron diffraction and the 9 T wide-angle chopper magnet for neutron spectroscopy developed by Oxford Instruments in collaboration with ISIS neutron source [3, 38, 40, 41].

The 14 T magnet for diffraction measurements is a high field, split pair magnet which consists of NbTi and Nb3Sn superconducting coils. Today the maximum magnetic field for a state-of-the-art split pair magnet is 15 T with an opening angle 3°, but by limiting the maximum field to 14 T, it is possible to increase the detector viewing angle to 7.5°. In order to optimise the magnet design for use on the WISH long-wavelength diffractometer, we have chosen an asymmetric split $-5°$ to $+10°$, which for a single crystal studies allows access to at least one extra scattering plane, apart from the main horizontal plane. The magnet split is supported by aluminium rings. The diffraction data demonstrate a very good angular coverage and data quality (low noise) thanks to the collimator and the cadmium shielding.

Similar to 14 T diffraction magnet, the 9 T wide-angle chopper magnet for spectrometry is a split pair magnet consisting of NbTi and Nb3Sn superconducting coils. As presented in Fig. 4.3, in this case the magnet split is supported by nonmagnetic stainless still wedges, rather than aluminium rings. This design allows to minimise the amount of the material in the beam. The chopper instruments derive their power from being able to survey large volumes of reciprocal space in a single measurement, and so it is vital to build a magnet that can exploit this advantage. This would necessitate a trade-off between maximum field and a wide aperture. Detailed modelling allowed us to design and manufacture 9 T magnet with 15 opening in vertical plane and 45 (two openings separated by 180) opening in horizontal plane. This magnet has been designed and built specifically for use on neutron spectrometers, such as MERLIN and LET (with the unprecedented continuous angular coverage of their detectors).

The 14 T and 9 T magnets share similar top parts of the recondensing cryostat with variable temperature inserts (VTI) but have different bottom parts accommodating split pair magnets and sample space. The design of recondensing magnet

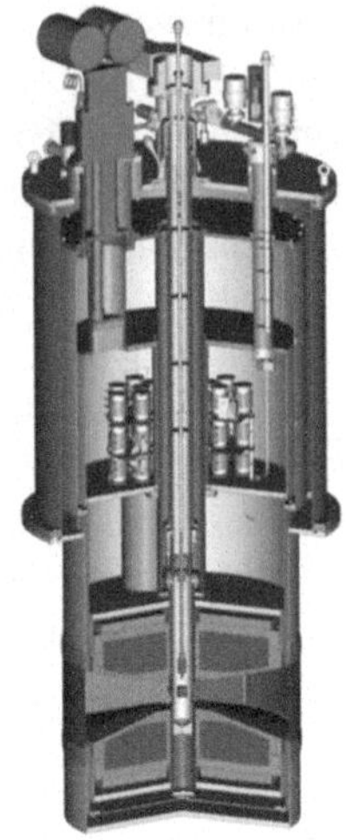

Fig. 4.3 Wide-angle chopper instrument 9 T superconducting magnet in the cryostat with helium recondensed by PTR

cryostats is usually based on the design of similar bath cryostats [12]. The superconducting magnet is immersed in the liquid helium. The radiation shield is cooled by the cooler's first stage, and the second stage recondenses helium directly in the helium vessel. Thus, the recondensing system does not consume any liquid helium during normal operations. The main advantage of this system is that all the magnet operating modes, for example cooling, running up to field and quenching, remain the same as for a standard magnet in a bath cryostat. This method also provides a homogeneous temperature distribution which is crucial for optimum magnet performance. Another significant advantage is the ability of the magnet to stay at field in the event of a power failure.

The cooling power for helium recondensing is provided by CryoMech pulse tube refrigerator PT410. This type of cryocoolers has no cold moving parts and, therefore, has low level of mechanical vibrations [12]. Additionally the pulse tube refrigerator requires little maintenance.

In a regime without VTI helium circulation and no ramping of filed pulse tubes on both magnets performs so efficiently that it recondenses all the helium in the cryostat and has an excess of cooling power between 600 and 650 mW [3]. This extra cooling power allows us to recondense the helium flow passing through a VTI during the operation of a dilution refrigerator insert [41].

4.5 Specialised Cryogen-Free Sample Environment Systems

In the last decade, cryogen-free technology has spread far beyond traditional low temperature/high magnetic field sample environment and has allowed conceptually new systems to be created. In this section I present a few examples of unconventional sample environments based on the cryogen-free approach.

There has been growing demand for investigations of the mechanical behaviour of materials at cryogenic temperatures, due to the recent progress in cryogenic technologies. Applications include cryogenic texture processing of alloys, strain sensitivity of superconducting magnet wires, cryogenic structural steels and low temperature shape memory alloys for space applications. ISIS has developed a cryogen-free cryogenic testing chamber for neutron scattering measurements of internal stress in engineering materials under loads up to 50 kN and at temperatures from 30 to 300 K [42]. Two CCRs provide the cooling power for keeping the sample at cryogenic temperatures. A similar approach has been used in the design of cryogenic load frames developed at the LANSCE [43] and JAEA [44] facilities. A distinctive characteristic of the JAEA device is the lower base temperature ~4.8 K, which allows the study of stress–strain effects in type 2 superconducting materials simultaneously with critical current measurements.

Recently, ISIS facility designed, manufactured and commissioned a new cryogenic testing chamber, for neutron scattering measurements of internal stresses in engineering materials under the load of up to 100 kN, in the temperature range from 6.5 to 300 K [45]. Complete cooling of the system from room to base temperature takes around 90 min. A photo of the experimental set-up is presented in Fig. 4.4. The cryostat design is similar to the one described in [42]. The main difference is the use of two-stage cryocoolers, instead of the single stage ones used in the previous design. The first stages of each of the two cryocoolers are thermally connected to the infrared radiation shield. They are also connected through copper braids to 100 K thermal anchoring points on the left and right grips. The sample holder grip is thermally linked to the second stage of the cryocooler. The design of left and right sides of the grips assembly is completely symmetric.

Excessive cooling power at first stages of cryocoolers allowed the ISIS team to incorporate a couple of high-temperature superconductor current leads into the system. This allows to supply a high current of up to 200 A to the sample, under

Fig. 4.4 Experimental set-up for the measurement of internal stress in engineering materials under loads. The uniaxial load up to 100 kN is provided by hydraulic loading rigs

load, simultaneously measuring stresses in the sample. In the system commissioning experiment, a dependence of the second-generation HTS tape sample, critical current has been successfully measured as a function of the sample's internal strain [46].

Another example of cryogenic sample environment is the cryogen-free cryostat, for cooling a 10 GPa Paris Edinburgh (PE) cell down to 3 K built by the ILL [9].

The 18 kg cell is attached to the second stage of the CCR and is enclosed in a tight chamber filled with exchange gas. It takes only 4 h to reach 3 K from room temperature. A cryogen-free cell based on a different design developed by FRMII allows high-pressure (up to 0.2 GPa) sample environment to operate in the temperature range of 20–600 K [47]. The cooling down time of this cell is similar to the ILL one.

Another version of cryogenic PE cell has been designed and developed at ISIS (please see photo in Fig. 4.5) [46]. This cryo-system is capable of varying the sample temperature over 17.5–300 K range in a little less than 3 h, cooling from room temperature to base temperature. The highest pressure achievable in the sample area is 30 GPa (300 kbar). The insert utilises two powerful 1.5 W at 4.2 K Sumitomo GM cryocoolers to cool just the sample and the standard profile tungsten carbide anvils. The variable temperature insert assembly is thermally insulated from the PE press. The temperature of the PE press cylinder housing is maintained close to ambient by means of a separate constant-temperature circuit. The system is designed to operate in completely automated regime, ran by a program script from the instrument computer.

Significant reductions in the neutron beam-time required for measurements with the new generation of neutron scattering instruments create the need for remote and fast sample handling to enable a high throughput of samples in a controlled

Fig. 4.5 Cryogenic PE cell based on two 1.5 W at 4.2 K Sumitomo GM cryocoolers

environment. A cryogen-free automated sample changer has been developed at SNS to serve this purpose [48]. The samples are contained in hermetically sealed vanadium cans attached to the sample handler. The system uses a CCR for cooling. The time to cool the sample from ambient temperature to ~10 K is approximately 7 min. A resistive heating coil can be used to extend the temperature range from 10 to ~350 K.

4.6 Cryogen-Free Technology in Neutron Scattering Instruments

The compactness and reliability of CCRs allowed them to occupy a new niche as coolers for components of neutron scattering instruments. Here we present a few examples of this approach.

The most widespread application of this kind is the CCR-based cryogenic beryllium filter. Polycrystalline beryllium Bragg cut-off occurs at an energy of 5.2 meV [46]. A block of beryllium transmits low energy neutrons up to this value, but scatters neutrons of higher energy. The beryllium filter based on this beryllium property is getting more and more popular in the design of neutron spectrometers. The transmission of the filter for neutrons of wavelengths beyond the cut-off depends mainly on phonon scattering. This can be largely removed by cooling the filter below 100 K and so removing the energy gain scattering from thermally excited phonons. Keeping beryllium filters at cryogenic temperature in the Tosca (ISIS) [5] and Vision (SNS) [49] instruments requires five CCRs, whereas in the case of the Lagrange (ILL) [50] and the recently upgraded Macs (NIST) [51] spectrometers, one or two CCRs provide enough cooling power.

Some instruments might require a cryogenic sample environment for the majority of experiments. In this case a cryogenic sample chamber can be incorporated in the design of the instrument. For the cryogenic chamber embedded in the Tosca instrument (ISIS), the necessary cooling power is provided by three CCRs, which allows measurements in the range between 3.7 K and room temperature [52].

A cryopump is a vacuum pump that traps gases and vapours by condensing them on a cold surface. In neutron instruments, cryopumps based on CCRs are used to improve the level of vacuum and accelerate pumping out of large vacuum chambers. For example, after venting the sample chamber of an SNS instrument Arcs, a combination of a large mechanical pump with a roots blower and a cryopump restores the vacuum to the 10-6 mbar range in under 15 min [53]. CCR-based cryopumps have also been used for the ISIS instruments Maps [54] and Merlin [6].

An elegant exploitation of superconductivity has been realised in the cryogen-free compact cryogenic polarisation analysis device (Cryopad) developed at the ILL [55]. This device takes advantage of Meissner shields made of niobium, to properly define the magnetic field and zero-field regions crossed by the incident and scattered neutron beams. Using the Cryopad, all the components of the complicated expression of the final polarisation vector of a neutron beam can be measured, which provides unique information on magnetic structures and nuclear/magnetic interferences occurring in the scattering process.

4.7 Conclusions

The cryogen-free revolution, driven by advances in cryocooler technology, started in the early 2000s and is not over yet. Every year, CCR-based technology conquers new areas traditionally occupied by conventional cryogenics. However, more than two decades of using cryocooler technology has allowed the accumulation of vast operational experience. Evaluation of the experience gained by ISIS, and other large neutron facilities, which are arguably the most intense users of CCR-based equipment, give us an opportunity to determine the advantages and disadvantages of these systems.

Advantages common for all CCR-based systems:

- Operational simplicity. Most of the cryogen-free systems only need maintenance services for the CCR cold-head (not required for PTRs) and the compressor after installation.
- Reduction or, in some cases, complete elimination of liquid cryogen top-ups, which is logistically demanding and expensive. CCR-based technology also reduces system downtime caused by top-ups.
- Significant reduction in technical resources. No need for specially trained personnel to prepare conventional cryostats for an experiment and perform top-ups.
- CCR technology is safer. The involvement of technical personnel regularly handling cryogens raises quite serious safety issues. The highest risk is associated with asphyxiation due to the lack of oxygen replaced by rapidly evaporating nitrogen or helium. This is a potentially lethal hazard. CCR technology significantly reduces, and in some cases completely eliminates all cryogenic hazards involved.
- Thermodynamic efficiency. Despite the fact that a cryocooler is less thermodynamically efficient than any industrial liquefier, the total efficiency of CCR-based systems is comparable with conventional cryogenic systems. This can be explained by the significant losses which conventional cryogenics experience due to transporting and storing cryogens and transferring them to cryostats. CCR technology eliminates these losses because cooling power is directly supplied from a cryocooler to the cooling apparatus.
- CCR technology significantly reduces the system size. The systems do not have a nitrogen vessel and do not require extra volume for holding liquid helium, which is necessary for providing sufficient hold time in the case of conventional cryostats.

- CCR technology is much more environmentally friendly than conventional cryogenics. Helium losses are minimised (and in some cases are completely eradicated) and the strategic gas can be saved for future generations. Evaluation of our operational experience also reveals some disadvantages of CCR-based systems.
- The most serious disadvantage of cryogen-free systems is the limited cooling power of the CCR. The most powerful GM cryocoolers and PTRs achieve no more than 1.5 W cooling power at 4.2 K. In conventional liquid helium-based systems, one can easily achieve more than an order of magnitude higher cooling power just by opening a needle-valve, but increasing helium consumption as a result. In comparison with conventional cryostat based on liquid cryogens, this means longer sample change time. However, in the case of recondensing systems, one can accelerate sample cooling by allowing helium boil-off.
- CCR-based equipment significantly increases demand for electricity and cooling water supplies, which in the case of massive scale operations can be a major issue.
- The cost of the cryocooler (tens of thousands dollars) can be a significant addition to the total cost of a cryogen-free cryostat.
- CCR operations generate significant noise, mechanical vibrations and magnetic field disturbances. This difficulty has been drastically alleviated by the absence of cold moving parts in the PTR design. However, in the cases of measurements extremely sensitive to such disturbances, special additional arrangements might be required.
- Some of the PTRs experience deterioration of cooling power (from 10% to 20% after ~5 years in operation), although the fast development in CCR technology promises significant improvement of PTR reliability in the near future.

CCR-based systems already make up a significant part of neutron facility cryogenic equipment, and the situation continues to change rapidly. The benefits of CCR technology lie not only in reducing helium consumption and eliminating the risks associated with global helium availability but also in the operational simplicity of cryogen-free systems, reduction in required technical resources and significantly improved safety. All these aspects are crucial for efficient exploitation of any large-scale neutron facility. A combination of CCR technology with advanced neutron optics, neutron scattering instrumentation and more powerful neutron sources opens up extraordinary new opportunities across broad areas of science.

Acknowledgements I am really grateful to Marsha Kirichek for proofreading this chapter's draft.

I also greatly appreciate the really fruitful collaboration with colleagues from HZB, ILL, Diamond, LLB, PSI, FRMII, B. I. Verkin Institute, RIKEN, Oxford Instruments and AS Scientific.

I would also like to thank all involved in the ISIS sample environment cryogenic development projects, particularly Z. A. Bowden, R. Down, B. Evans, Jeff Keeping, Rob Major as well as the entire ISIS sample environment cryogenic section.

References

1. Bailey IF (2003) A review of sample environments in neutron scattering. Z Kristallogr 218:84
2. Kirichek O (2012) Impact of the cryogen free revolution on neutron scattering laboratories. Mod Phys Lett B 26:1230006
3. Kirichek O, Brown J, Adroja DT, Manuel P, Kouzmenko G, Bewley RI, Wotherspoon R (2012) New generation of cryogen free advanced superconducting magnets for neutron scattering experiments. J Phys Conf Ser 400:052013
4. Windsor CG (1981) Pulsed neutron scattering. Taylor & Francis LTD, London, p 357
5. Colognesi D, Celli M, Cilloco E, Newport RJ, Parker SF, Rossi-Albertini V, Sacchetti F, Tomkinson J, Zoppi M (2002) TOSCA neutron spectrometer: the final configuration. Appl Phys A 74:S64
6. Bewley RI, Eccleston RS, McEwen KA, Hayden SM, Dove MT, Bennington SM, Treadgold JR, Coleman RLS (2006) MERLIN, a new high count rate spectrometer at ISIS. Phys B 385–386:1029
7. Feder T (2005) Government handling of helium gets report card: think again. Phys Dent Today 63(2010):28
8. Simkin VG (2001) Sample environment around the IBR-2 spectrometers. JINR Commun: 23
9. Bourgeat-Lami E, Chapuis JF, Chastagnier J, Demas S, Gonzales JP, Keay MP, Laborier JL, Lelievre-Berna E, Losserand O, Martin P, Melesi L, Petoukhov A, Pujol S, Ragazzoni JL, Thomas F, Tonon X (2006) Overview of the projects recently developed by the advanced neutron environment team at the ILL. Physica B 385–386:1303
10. Evans BE, Down RBE, Keeping J, Kirichek O, Bowden ZA (2008) Cryogen-free low temperature sample environment for neutron scattering based on pulse tube refrigeration. Meas Sci Technol 19:034018
11. Kirichek O, Evans BE, Down RBE, Bowden ZA (2009) Cryogen free low temperature sample environment for neutron scattering experiments. J Phys Conf Ser 150:012022
12. Kirichek O, Carr P, Johnson C, Atrey M (2005) Nuclear magnetic resonance magnet actively cooled by pulse tube refrigerator. Rev Sci Instrum 76:055104
13. Tomaru T, Suzuki T, Haruyama T, Shintomi T, Yamamoto A, Koyama T, Li R (2004) Vibration analysis of Cryocoolers. Cryogenics 44:309
14. Wang C (2001) Helium liquefaction with a 4 K pulse tube cryocooler. Cryogenics 41:491
15. Will ME, Tanaeva IA, Li R, de Waele ATAM (2004) New rotary valves for pulse-tube refrigerators. Cryogenics 44:793
16. Brochier D (1977) ILL tech report 77/74
17. Kirichek O, Down RBE, Keeping J, Evans BE, Bowden ZA (2012) Cryogen free sample environment for neutron scattering experiments at ISIS. J Phys Conf Ser 340:012009
18. Chapman CR, Evans BE, Dudman MP, Keeping J, Down RBE, Kirichek O, Bowden ZA (2011) Cryogen-free cryostat for neutron scattering sample environment. Cryogenics 51:146
19. Kirichek O, Foster J, Down RBE, Clapton D, Chapman CR, Garside J, Bowden ZA (2013) Top loading cryogen free cryostat for low temperature sample environment. J Low Temp Phys 171:737
20. Frossati G, Godfrin H, Hebral B, Schumacher G, Thoulouze D (1978) Proceedings of the ultralow temperatures symposium, Hakone, Japan (1977)
21. Frossati G (1978) J Phys Coll 39:1578
22. Bradley DI, Bradshaw TW, Guenault AW, Keith V, Locke-Scobie BG, Miller IE, Pickett GR, Pratt WP Jr (1982) A dilution refrigerator combining low base temperature, high cooling power and low heat leak for use with nuclear cooling. Cryogenics 22:296
23. Koike Y, Morii Y, Igarashi T, Kubota M, Hiresaki Y, Tanida K (1999) A dilution refrigerator using the pulse tube and GM hybrid cryocooler for neutron scattering. Cryogenics 39:579
24. Uhlig K (2002) 3He/4He dilution refrigerator with pulse-tube refrigerator precooling. Cryogenics 42:73
25. Uhlig K (2004) "Dry" dilution refrigerator with pulse-tube precooling. Cryogenics 44:53

26. Uhlig K (2008) 3He/4He dilution refrigerator with high cooling capacity and direct pulse tube pre-cooling. Cryogenics 48:511
27. Diallo SO, Azuah RT, Kirichek O, Taylor JW, Glyde HR (2009) Limits on Bose-Einstein condensation in confined solid 4He. Phys Rev B 80:060504
28. Glyde HR, Diallo SO, Azuah RT, Kirichek O, Taylor JW (2011) Bose-Einstein condensation in liquid 4He under pressure. Phys Rev B 83:100507
29. Glyde HR, Diallo SO, Azuah RT, Kirichek O, Taylor JW (2011) Atomic momentum distribution and Bose-Einstein condensation in liquid 4He under pressure. Phys Rev B 84:184506
30. Nucciotti A, Schaeffer D, Alessandria F, Ardito R, Barucci M, Risegari L, Ventura G, Bucci C, Frossati G, Olcese M, de Waard A (2008) Design of the cryogen-free cryogenic system for the CUORE experiment. Low Temp Phys 151:662
31. Prouve T, Godfrin H, Gianse C, Triqueneaux S, Ravex A (2007) Pulse-tube dilution refrigeration below 10 mK for astrophysics. J Low Temp Phys 148:909
32. Batey G, Buehler M, Cuthbert M, Foster T, Matthews AJ, Teleberg G, Twin A (2009) Integration of superconducting magnets with cryogen-free dilution refrigerator systems. Cryogenics 49:727
33. Yayama H, Yoshimura M (2009) Installation of a superconducting magnet in a cryogen-free dilution refrigerator. J Phys Conf Ser 150:012056
34. Cho A (2009) Helium-3 shortage could put freeze on low-temperature research. Science 326:778
35. Meissner M, Smeibidl P (2001) Neutron scattering at BENSC under extreme conditions: up to 17 tesla and down to 25mK. Neutron News 12:12
36. Katano S, Minakawa N, Metoki N, Osakabe T, Suzuki J, Koike Y, Ishii Y (2002) Liquid-He-free 10T superconducting magnet for neutron scattering. Appl. Phys A 74:S270
37. Pooke DM, Chamritski V, Fee M, Gibson S, King BT, Tallon JL, Meissner M, Feyerherm R, Olsen SR, Kennedy SJ, Robinson RA (2009) HTS 5 Tesla synchrotron and neutron beamline magnets. IEEE Trans Appl Supercond 19:1372
38. Down RBE, Kouzmenko G, Kirichek O, Wotherspoon R, Brown J, Bowden ZA (2010) Cryogen free high magnetic field sample environment for neutron scattering. J Phys: Conf Ser 251:012092
39. Lelievre-Berna E, Brown J, Jones H, Kousmenko G, Losserand O, Pickering P, Tonon X, Turc S (2010) ILL preprint
40. Brown FJ (2010) Aspects of superconducting magnet design for neutron scattering sample environments. J Phys Conf Ser 251:012093
41. Kirichek O, Down RBE, Kouzmenko G, Keeping J, Bunce D, Wotherspoon R, Bowden ZA (2010) Operation of superconducting magnet with dilution refrigerator insert in zero boil-off regime. Cryogenics 50:666
42. Oliver E, Evans BE, Chowdhury M, Major R, Kirichek O, Bowden A (2008) Novel testing chamber for neutron scattering measurements of bulk stress in engineering components at cryogenic temperatures. Meas Sci Technol 19:034019
43. Tao K, Wall JJ, Li H, Brown DW, Vogel SC, Choo H (2006) In situ neutron diffraction study of grain-orientation-dependent phase transformation in 304L stainless steel at a cryogenic temperature. J Appl Phys 100:123515
44. Tsuchiya Y, Suzuki H, Umeno T, Machiya S, Osamura K (2010) Development of a cryogenic load frame for a neutron diffractometer. Meas Sci Technol 21:025904
45. Kirichek O, Timms JD, Kelleher JF, Down RBE, Offer CD, Kabra S, Zhang SY (2017) Sample environment for neutron scattering measurements of internal stresses in engineering materials in the temperature range of 6 K to 300K. Rev Sci Instrum 88:025103
46. Kirichek O (2017) Sample environment for neutron scattering experiments at ISIS. J Neutron Res 19:57–63
47. FRM II Experimental Facilities Booklet 2010. http://cdn.frm2.tum.de/fileadmin/stuff/instruments/BlueBook/exp-fac_weboptimiert-2p.pdf

48. Rix JE, Weber JKR, Santodonato LJ, Hill B, Walker LM, McPherson R, Wen-zei J, Hammons SE, Hodges J, Rennich M, Volin KJ (2007) Automated sample exchange and tracking system for neutron research at cryogenic temperatures. Rev Sci Instrum 78:013907
49. Seeger PA, Daemen LL, Larese JZ (2009) Resolution of VISION, a crystal-analyzer spectrometer. Nucl Instr Meth Phys Res Sec A 604:719
50. Harrison A, Mart'nez JL, Wagner R (2010) Renaissance: a decade of development at ILL. Neutron News 21:11
51. Rodrigues JA, Adler DM, Brand PC, Broholm C, Cook JC, Brocker C, Hammond R, Huang Z, Hundertmark P, Lynn JW, Maliszewskyi NC, Moyer J, Orndorff J, Pierce D, Pike TD, Scharfstein G, Smee SA, Vilaseca R (2008) MACS-A new high intensity cold neutron spectrometer at NIST. Meas Sci Technol 19:034023
52. Down RBE, Ramirez-Cuesta AJ, Major RA, Keeping J, Rudić S, Kirichek O (2014) Cryogenic sample environment on TOSCA. J Phys Conf Ser 554:012007
53. Abernathy DL, Stone MB, Loguillo MJ, Lucas MS, Delaire O, Tang X, Lin JYY, Fultz B (2012) Design and operation of the wide angular-range chopper spectrometer ARCS at the Spallation Neutron Source. Rev Sci Instrum 83:015114
54. Perring TG, Taylor AD, Osborn R, Mc Paul D, Boothroyd AT, Aeppli G (1994) Proceedings of ICANS XII I-60; RAL report 94-025
55. Lelivre-Berna E, Bourgeat-Lami E, Gibert Y, kernavanois N, Locatelli J, Mary T, Pastrello G, Petukhov A, Pujol S, Rouques R (2005) Advances in spherical neutron polarimetry with Cryopad. Physica B 356:141

Chapter 5
Integration of Adiabatic Demagnetization Refrigerators with Spaceflight Cryocoolers

Peter J. Shirron and Michael J. DiPirro

Abstract Cryogenic cooling is an increasingly vital technology for ultra-high-resolution space-based instruments. The vast majority of detectors for X-ray, infrared, and sub-millimeter radiation now rely on operating at very low temperature to achieve the sensitivities and low signal background required for the upcoming and future missions. Operating temperatures in the 50–100 mK range have become common, and some new detector technologies can benefit from even lower colder operation. Among the refrigeration techniques that can achieve such temperatures, adiabatic demagnetization refrigerators (ADR) have many advantages for space missions, including high efficiency, lack of gravity dependence, and wide operating range. However, as an inherently single-shot type of cooler, and one that requires relatively high current to drive the magnetic cycle, there are challenges for their implementation in space instruments and coupling to the cryocoolers and cryogenic systems that support their operation. In this chapter, we discuss these challenges, how they affect ADR and system design and operation, and options for optimizing performance and expanding capabilities to meet the demands of future space instruments.

5.1 Introduction

Space offers an unparalleled vantage point for a wide range of astronomical and earth observations. Free from atmospheric distortion and seismic disturbances, orbiting satellites are able to resolve far fainter signals than is possible from terrestrial platforms. This has driven the development of extremely sensitive detectors (bolometers for infrared to submillimeter wavelengths and microcalorimeters [1] for X-rays) that achieve superior resolution in part by operating at very low temperature.

P. J. Shirron (✉) · M. J. DiPirro
NASA/Goddard Space Flight Center, Cryogenics and Fluids Branch, Code 552, Greenbelt, MD, USA
e-mail: peter.shirron@nasa.gov

M. D. Atrey (ed.), *Cryocoolers*, International Cryogenics Monograph Series,
https://doi.org/10.1007/978-3-030-11307-0_5

The first instruments used detectors cooled to about 1 K, but eventually the push for higher sensitivity resulted in the implementations that depended on cooling into the deep sub-kelvin range. Today, detector technologies [2–4] and multiplexed readout schemes [5–8] have matured to the point that future space missions, for example, the Athena [9] X-ray observatory can propose the use of large arrays, in the kilo- to megapixel range.

The result is that there has been a steady evolution in cryogenic architectures, and the cooling capabilities of cryogenic cooling systems developed to support these missions. The first astronomical observations using low-temperature detectors were made aboard the Kuiper Airborne Observatory (KAO) [10] using bolometers cooled by superfluid helium. Since the KAO instruments were operating in 1-g, traditional dewars could be used. But the immediate success of those observations provided a push for applying the new detector technologies to orbital platforms, for which the main technical issue for the cryogenic system was how to contain liquid helium in 0-g.

The successes of [IRAS, COBE] paved the way for the development of sub-kelvin coolers—principally drawing from the established technologies of 3-helium sorption, dilution refrigeration, and adiabatic demagnetization refrigeration (ADR), which could take advantage of a 1 K heat sink. These were all successfully implemented in space in the mid-2000s: Herschel [11] and Planck [12] were co-launched in 2006 and, respectively, used a 3-helium sorption refrigerator [13] operating at 300 mK and an open-cycle dilution refrigerator [14] operating at 100 mK, and Astro-E2/Suzaku [15] launched in 2005 carrying an single-stage ADR [16] operated at 60 mK. It should be noted that while both sub-kelvin Herschel and Suzaku refrigerators used superfluid helium at $\sim$1 K as their heat sink, the Planck dilution cooler based its operation of a 4 K cryocooler [12].

As these missions were being implemented, significant investments were being made in space cryocooler technologies, capable of operating at 5 K and below, within NASA, ESA, and JAXA. The primary goal was to overcome limitations on lifetime inherent to the use of stored cryogens, although they would impose their own limitations on the range of temperatures and cooling powers produced. NASA efforts targeted 4–6 K operation, with a target cooling power of 40 mW, with the expectation that concurrent developments in sub-kelvin technologies would enable their operation from such temperatures. JAXA, on the other hand, supported the efforts to develop both 1 K- and 4 K-class Joule-Thomson cryocoolers [17] and by early 2000s had working prototypes. Cooling powers were similarly limited to 10 and 40 mW, respectively.

At the same time, the advances in detector arraying capability were imposing more stringent cooling requirements that in some cases were beyond the capability of single-stage ADRs, especially given the higher cryocooler heat sink temperature.

This made it necessary to consider multistage designs and architectures that could expand the performance in one or more of the following ways:

	Goal or typical value
Wider operating temperature range	>4.5 K
Higher cooling power	>5 μW at 50 mK
Multiple, stable operating temperatures, continuous operation	50–100 mK, 1–2 K

Having multiple operating temperatures is beneficial for several reasons. In some cases, an intermediate temperature is required for cooling the instrument components or amplifiers, such as SQUIDs. But even when it is not required, having one or more intermediate temperatures enables conducted or dissipated heat loads to be intercepted, thereby reducing the parasitic and dissipated loads at the coldest stage. This allows more of the cooling capacity to be used for instrument loads, thereby increasing the refrigerator's efficiency, cooling power, and operating temperature range.

Continuous operation is also an effective strategy for improving cooling power and operating range, and it has important advantages for coupling a sub-kelvin cooler, such as an ADR, to a cryocooler. The main drawback of single-shot operation is that it pushes the system design toward one in which the sub-kelvin cooler can maintain a low temperature for an extended duration, typically 24–48 h, and can regenerate itself in a relatively short period, typically 1 h. With this capability, an instrument can be operational for more than 95% of the time. But the disadvantages are that (1) the cooling power at low temperature is extremely limited, with potential impacts on detector or instrument capability, and (2) the cryocooler heat sink is subjected to a widely varying heat flux. The latter can impact the design or operation of the cryocooler, but more importantly it requires oversizing the cryocooler based on the peak heat flux rather than the average.

Continuous operation alleviates the constraint of concentrated heat rejection by allowing heat to be rejected much more frequently and over a larger fraction of the refrigeration cycle. The result is a reduction in peak heat rejection rate to the cryocooler, allowing the cryocooler to be sized closer to the average heat load, with a corresponding reduction in input power. Continuous operation also increases the ADR's cooling power per unit mass, allowing it to meet more demanding cooling requirements within the constraints of the existing spaceflight cryocoolers.

But despite the ability to produce continuous cooling, each ADR stage in such a system still operates in an inherently single-shot manner, cycling individually to absorb heat for a period of time at low temperature and reject heat for a period of time at high temperature. This results in a heat load to the cryocooler heat sink that varies significantly over time, requiring a cooling power capability that may be considerably higher than the average load. Given the relatively limited cooling power of space cryocoolers, this creates the single greatest challenge for space instruments.

Nevertheless, as will be discussed, the evolution of multistage architectures and cycling strategies can relieve some of these constraints and enable ADR systems to be well-matched to the operating characteristics of space cryocoolers. Although more complex, these ADRs result in higher system efficiency and reduce the cryogenic system's burden on spacecraft resources.

5.2 ADR Design and Thermodynamic Considerations

In recent years, ADR design has evolved from single-stage systems that provide a single operating temperature using a fixed temperature heat sink to multistage designs [18] that can have multiple operating temperatures and operate continuously. But regardless of how stages are present or how they are configured, each operates on the same principles and has the same basic elements.

5.2.1 *Basis of ADR Operation*

The refrigeration cycle is based on the magnetocaloric effect (MCE), in which a decreasing (or increasing) magnetic field in the volume of a magnetic refrigerant produces a decreasing (or increasing) temperature and/or the absorption (or rejection) of heat. The MCE is a quantum phenomenon which occurs in materials containing magnetic atoms having an odd half-integer spin angular momentum, J, where the $2J + 1$ states corresponding to the allowed values of the z-component of spin, $J_z = +J/2, +J/2 - 1, -J/2$, are degenerate. In the absence of any external magnetic field, all $2J + 1$ states have equal occupation, and the entropy per molecule is $S_{\max} = k_{\mathrm{B}}\ln(2J + 1)$, where k_{B} is Boltzmann's constant. In the presence of an external field, B, occupation is biased toward those spin states that are aligned with the field, as opposed to anti-aligned, and toward the largest J_z value. Reducing the effective number of states reduces the entropy, requiring the expulsion of heat, $\Delta Q = T\Delta S$. In the limiting case of very high field and/or low temperature, i.e., $BJ \gg k_{\mathrm{B}}T$, only the $J_z = +J$ state is energetically favorable, and the corresponding entropy, $k_{\mathrm{B}}\ln(1)$, is suppressed to zero.

In the general formulation [19], the entropy per mole of a system of noninteracting spins as a function of temperature, T, and applied field, B, is given by

$$\frac{s\,(B,T)}{R} = x\coth(x) - (2J+1)\,x\coth\left((2J+1)\,x\right) + \ln\left(\frac{\sinh\left((2J+1)\,x\right)}{\sinh(x)}\right) \tag{5.1}$$

$$x \equiv \frac{\mu_{\mathrm{B}} g B}{2k_{\mathrm{B}}T} = \frac{0.336\,(K/T)\,gB}{T}$$

where μ_{B} is the Bohr magneton, and g is the Land g-factor for the magnetic ion (usually 2).

In real materials, the spins are rarely so dilute that they can be considered noninteracting. Instead internal interactions cause the spins to progressively self-align as temperature is lowered. The resulting entropy suppression depends on the density of magnetic ions and magnetic moment (J). To represent these internal interactions, it is common to introduce a background field [20, 21], b, which is added to the applied field in quadrature to yield an effective field $B_{\mathrm{eff}} = (B^2 + b^2)^{1/2}$, that is used in Eq (5.1) in place of B.

While these non-idealities and their effect on magnetic properties are important when designing and optimizing ADR stages, they have little effect on parameters such as heat output and thermodynamic efficiency that we are concerned with here. It is therefore sufficient to treat the refrigerant as a lumped mass whose entropy can be manipulated by an applied magnetic field to achieve a simple and efficient refrigeration cycle. The keys to the simplicity and high efficiency of the ADR cycle are, first, that the entropy as given by Eq. (5.1) depends only on the ratio of B_{eff}/T, such that under adiabatic conditions, i.e., $S(B, T) = \mathrm{constant}$, the salt pill's temperature varies linearly with B_{eff}, and second, that changes in magnetic field and temperature are reversible.

An ADR stage consists of the following basic components: a capsule of magnetic refrigerant (called a salt pill), a superconducting magnet, a heat switch, and, typically, a suspension system to physically support the salt pill within the magnet. The latter may not be necessary depending on the application. The heat switch functions to thermally connect or disconnect the salt pill to a heat sink, which may variously be another stage of refrigeration, a liquid helium bath, or a mechanical cryocooler. Figure 5.1 is a schematic representation of this configuration.

The ADR cycle consists of four discrete steps: (1) adiabatic magnetization of the salt pill until warms it above the heat sink, at which point the heat switch is closed; (2) isothermal magnetization to full field as heat flows to the sink, at which point the heat switch is opened; (3) adiabatic demagnetizion of the salt pill until it reaches the desired operating temperature, and (4) isothermal demagnetizion as the stage absorbs heat from its load. When the field reaches a minimum value (usually zero), the cycle is repeated.

The salt pill's cooling capacity at its operating temperature, T_{low}, is $\Delta Q = n \cdot \varepsilon \cdot T_{\mathrm{low}}[s(B_{\mathrm{min}}, T_{\mathrm{low}}) - s(B_{\mathrm{max}}, T_{\mathrm{high}})]$, where n is the number of moles of refrigerant, ε is the heat absorption efficiency, B_{max} and B_{min} are the maximum and minimum magnetic fields that define the ends of the isothermal magnetization

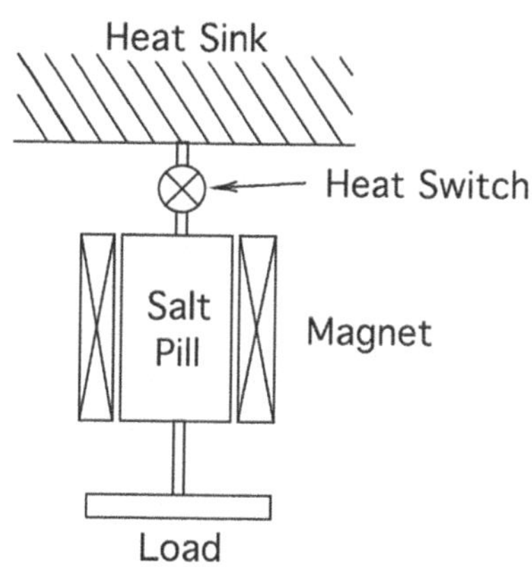

Fig. 5.1 Schematic of a single-stage ADR. Not shown is the suspension system that physically supports the salt pill in the magnet. The magnet is usually thermally connected to the heat sink

and demagnetization steps, and T_{high} is the temperature from which the salt pill is demagnetized (usually only slightly higher than the heat sink temperature, T_{sink}).

Although there is no intrinsic limit on T_{high}, there are fundamental constraints based on both magnet technologies and the properties of the magnetocaloric refrigerants. The primary limitation is the ability to create the magnetic fields needed to achieve reasonable temperature suppression during demagnetization. Progressively higher field is needed as T_{high} increases, at the same time that the critical field of superconductors decreases. For NbTi magnets, the crossover occurs at about 6 K. Advances in Nb_3Sn [22], MgB_2, and HTS [19] magnets are beginning to accumulate to the point that producing the fields required for operation at higher than 6 K, and especially with modest currents (<10 A), will soon be practical. But even given this capability, at elevated temperature, the lattice heat capacity of the refrigerant becomes competitive with the magnetic component. The temperature change associated with demagnetization then becomes too small compared to absolute temperature to form the basis for practical refrigeration, and one must instead consider such techniques as active magnetic regenerative refrigeration [23] becomes more advantageous.

5.2.2 Multistaging to Increase the Temperature Range and Performance

The operation of a single ADR stage requires the salt pill to demagnetize from the heat sink temperature to the operating point. The heat loads that the salt pill must absorb come from a number of sources, the heat switch, the suspension system, sensor wiring, and the instrument being cooled, all of which are thermal conductors between the salt pill and heat sink. In early ADRs, the relevant temperature span was typically ~1.5 to 50–60 mK. (Thermal radiation from 1.5 K is negligible.)

As space cryocoolers were being developed, the need for ADRs operating from heat sinks in the 4–5 K range was foreseen. For a single-stage ADR, the parasitic heat loads would be dramatically increased, to tens of W. As a mitigation, first a 2-stage [24] concept was developed, and later this was extended to multiple stages [25]. The main advantages of using two or more stages to span a given temperature range are as follows. First, the upper stage(s) act as heat intercepts for the coldest stage, in that components attached to that stage can be anchored at their warm end to other ADR stages rather than the heat sink. Second, the upper stage(s) can be used to pre-cool the colder stages, so that they can be demagnetized from a lower temperature than the heat sink, significantly reducing the magnetic fields required. A third advantage is that the upper stages, since they do not need to cool nearly as deeply as the coldest stage, can use different refrigerants with much higher density and cooling capacities. The combination of these effects dramatically improves the system's cooling power and reduces its mass compared to a single-stage ADR.

The first ADR launched into space, aboard Astro-E2/Suzaku, used a 1.2 K superfluid helium heat sink, and consisted of a single stage using a 920 g ferric

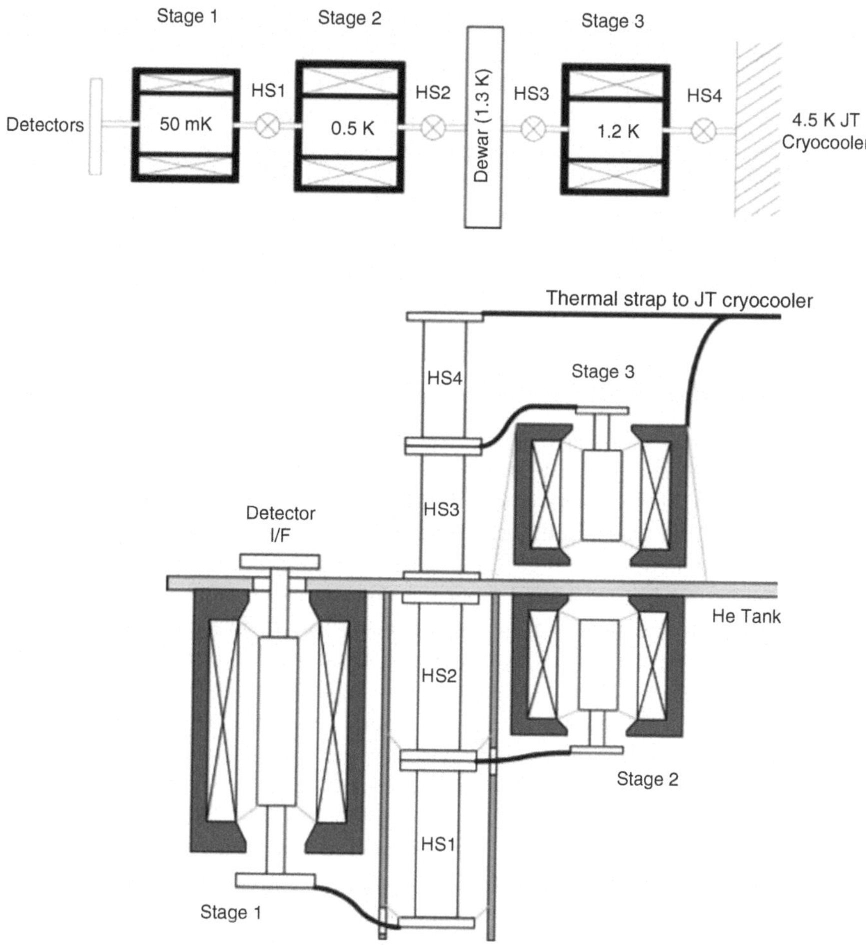

Fig. 5.2 Schematic of the Astro-H 3-stage ADR, and the physical and thermal arrangement of the stages within the cryogenic system. Components shown in yellow represent thermally isolating structural supports, such as the Kevlar suspension systems [28]. The heavy black lines represent thermal straps between stages and between the ADR and JT cryocooler. The assembly is mounted to a tank initially containing liquid helium

ammonium alum (FAA) salt pill in a 2 T magnet. When Astro-H (originally called the Next X-ray Telescope, or NeXT) was proposed in 2007, the instrument concept was to augment a superfluid helium tank with a 1.8 K JT cryocooler, both to reduce the heat load on the helium and to provide a redundant heat sink. The optimal ADR design was a 2-stage ADR directly coupled to the helium tank and coupled to the cryocooler through a gas-gap heat switch. Later, the cryocooler was changed to a 4.5 K version, and the ADR, shown schematically in Fig. 5.2, was modified to add a third stage that could bridge the larger temperature range [26, 27].

The Astro-H ADR has two operating modes. When liquid helium is present, stages 1 and 2 are cycled, rejecting heat to the liquid helium and then cooled to 50 mK and 0.5 K, respectively. When the liquid is exhausted, stages 2 and 3 are operated as a continuous ADR to cool the empty helium tank to 1.5 K using the 4.5 K JT cryocooler as a heat sink. Stage 1 is operated essentially as a single-stage ADR, rejecting heat to ~1.5 K, and then cooling to 50 mK.

As this was the first combination of an ADR operating from a space cryocooler, it offers unique insights into issues related to the integration and operation of the two cryogenic systems.

5.2.3 *Heat Rejection from an ADR*

One of the fundamental issues to be addressed in coupling ADRs to a cryocooler is the way in which ADRs generate heat. The heat generated by an ADR comes from three major sources: the salt pill, dissipation in the heat switch, and hysteresis in the magnet (plus magnetic shielding, if any is used). While the heat rejected from the salt pill can be readily calculated from the entropy expression given previously, the other two terms depend on the particular choices of heat switch technology and configuration of the magnet and magnetic shielding. Nevertheless, the relative magnitudes of these heat loads appear to follow general trends, as evident from the following examples of both spaceflight and laboratory ADRs, which allows the heat loads to be estimated for a generalized system.

Table 5.1 lists the top-level design parameters for the single-stage Astro-E2 ADR, the 3-stage Astro-H ADR, and a 4-stage continuous ADR (CADR). For Astro-H, stages 1 and 2 are grouped as a unit, as the manner in which they are cycled prevents a determination of individual contributions to the heat output. Stage 3's heat output, however, operates independently, allowing its heat load to be tabulated separately. The refrigerant materials referenced are ferric ammonium alum, or FAA, chromic potassium alum, or CPA, and gadolinium lithium fluoride, or GLF.

Table 5.1 Design parameters of selected ADR systems

ADR system	Refrigerant	Magnetic field (T)	Heat absorption temperature (K)	Heat rejection temperature (K)
Astro-E2	920 g FAA	2.0	0.06	1.3
Astro-H, stages 1, 2	270 g	2.0	0.05	1.2
	150 g GLF	3.0	0.50	1.2
Astro-H, stage 3, 4-stage laboratory, continuous ADR	150 g GLF	3.0	1.2	4.5
	60 g CPA	0.1	0.05	0.05
	100 g CPA	0.8	0.045	0.3
	100 g CPA	1.6	0.27	1.5
	65 g GLF	4.0	1.3	4.5

Table 5.2 Design parameters of the selected ADR systems

ADR system	Slate pill heat (J)	Hysteresis heat (J)	Heat switch dissipation (J)	Total heat per cycle (J)	Salt pill heat as fraction of total (%)
Astro-E2	14.76	6.58	3.69	25.03	59.1
Astro-H, stages 1, 2	7.67	4.33	0.79	12.79	60.0
Astro-H, stage 3, 4-stage laboratory, continuous ADR	17.62	1.85	0.39	12.79	77.3
	4.70	1.30	0.00	6.00	78.3

Table 5.2 gives the heat output contributions for those systems. It should be noted that the magnet built for Astro-E2 was actively shielded by means of a bucking coil that minimized the magnetic field at the detector array location, where both the Astro-H and 4-stage CADR used ferromagnetic shields on each magnet. While hysteresis in superconducting coils can be calculated [29] based on the details of the wire used and coil geometry, specialized magnetic field modeling software is required when using combinations of superconducting coils and ferromagnetic shielding. For this reason, we present the hysteresis only as an aggregate for each magnet or magnet/shield combination. We have not observed any rate dependence to the hysteresis heating and have concluded that at (de)magnetization rates used in each cycle are slow enough to reduce eddy current heating in the coils, magnet mandrels and magnetic shielding to at most a very small fraction of the total.

It is reasonable to question whether these examples are too specific to their applications and therefore not be representative of ADRs designed for different cooling capacities or heat reject temperatures. We contend that they are, and to support this, we note that most ADR design parameters scale linearly with cooling capacity and reject temperature. For example, designing for a different cooling capacity involves proportionally scaling the volume of the salt and, consequently, the magnet, yielding a proportional change in salt pill and hysteresis heat loads. For a fixed heat switch conductance, the time required to reject heat from the salt pill, during which the heat switch is powered, also increases proportionally. Thus the result is an essentially proportional change in heat rejected from the salt pill, magnet, and heat switch.

Increasing (or decreasing) the reject temperature proportionally changes the amount of heat rejected from the salt pill and requires a proportionally higher (or lower) magnetic field to achieve the same entropy capacity, with a proportional change in hysteresis heat generation. And, as above, the heat switch dissipation scales with salt pill heat output, yielding the same linear scaling of heat loads with reject temperature as with cooling power.

This argument breaks down when the change in operating and/or reject temperatures enables a different magnetic refrigerant to be used [30]. ADRs operating at below about 0.3 K are constrained to use low-density hydrated salts, whereas higher operating points enable the use of much higher density rare earth materials. The higher density yields higher cooling power per volume, thereby increasing the heat output of the salt pill relative to the magnet's hysteresis.

In Table 5.2, it can be seen that heat loads scale roughly proportionally with the mass of refrigerant, whether the mass is contained in a single stage or distributed among more than one. What we wish to emphasize, though, is, first, that the heat load arising from the salt pill(s) is the largest contribution and, second, that the fraction that it represents of the total is similar for ADRs that reject heat at similar temperature. The Astro-E2 and Astro-H 2-stage ADRs both reject heat to ~1.2–1.3 K, and the salt pill contribution is about 60% of the total. In the 4-stage CADR, all heat produced by the salt pills is cascaded up to the fourth stage and then transferred to the heat sink. That is, regardless of whether there are internal transfers, one needs to consider only the temperature at which heat flows out of the ADR, which for the 4-stage CADR and Astro-H stage 3 is ~4.5 K. The salt pill heat output is similarly about 80% of the total.

Assuming this analysis can be applied to a generalized system, it provides a means of estimating the total heat output once the basic design details (mass and type of refrigerant, and magnetic field for each stage) are known.

5.2.4 Phasing of Heat Loads

An important distinction among the heat loads, though, is their phasing. Figure 5.3 presents the time history of heat flow from the Astro-H ADR during a recycling operation, with the elapsed time starting when the 50 mK stage reaches the end of its hold time. (Only stages 1 and 2 are being operated, and their heat sink is a liquid helium bath at 1.2 K.) Magnetic hysteresis heat is generated as the stages are magnetized at their maximum rate to the heat sink temperature. The heat switch is then powered on, giving rise to a heat flow from the salt pill, which dominates until peak field is reached and the heat switch is powered off. Further hysteresis heat is generated as the ADR is demagnetized to its operating temperature(s).

Heat flow is highly variable during the recycling process, with a broad peak occurring during isothermal magnetization and smaller peaks during adiabatic magnetization and demagnetization. When heat is being rejected to a helium bath, as in this case, the variability over time is not particularly relevant as helium is evaporated simply at a rate proportional to the instantaneous load. However, when stage 3 of the Astro-H ADR is operated, to lift heat from the helium tank to the 4.5 K cryocooler, this variability, shown in Fig. 5.4, can be problematic.

The peak heat flow can be controlled by regulating the thermal gradient between the ADR stage and cryocooler, based on the thermal conductance of the heat switch and thermal links between the two. There are inherent advantages to imposing the highest possible heat flux, as it results in the shortest recycle time and either the highest possible cooling power for continuous ADRs or the highest duty cycle for single-shot systems. Commercial cryocoolers have cooling powers well in excess of the heat load that a practical ADR can produce, but for space cryocoolers, the

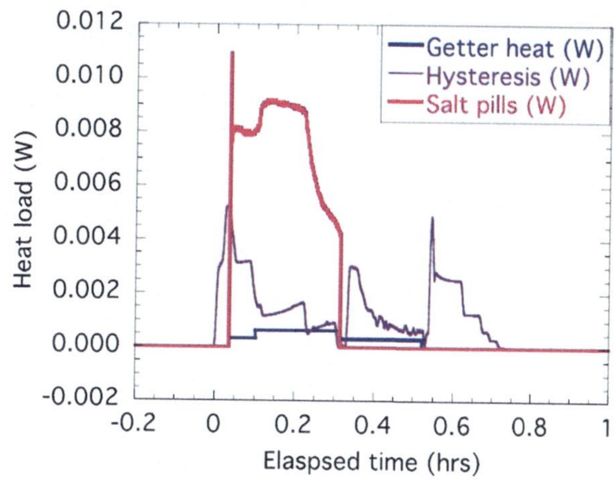

Fig. 5.3 Heat output from the Astro-H ADR (salt pill heat, hysteresis, and heat switch dissipation) during a recycle operation when stages 1 and 2 reject heat to a 1.2 K helium bath

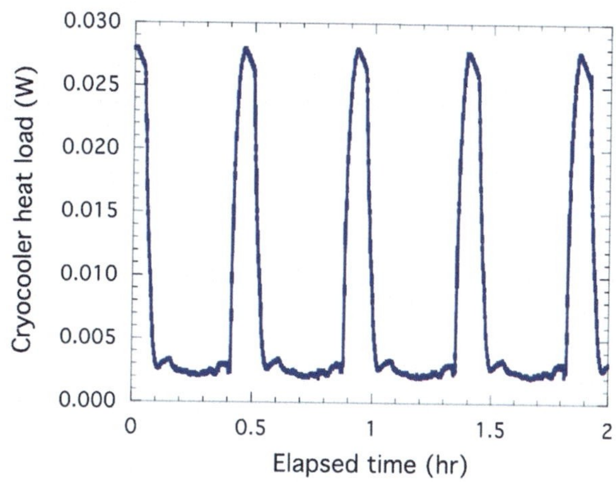

Fig. 5.4 The periodic heat rejection rate from the third stage of the Astro-H ADR as it cycles between cooling at 1.2 K and rejecting heat to 4.5 K

available cooling power is the limiting factor in how rapidly a stage can be recycled. Figure 5.5 shows the performance of two representative space flight coolers. The first is the measured cooling power versus precooling temperature for the 4 K JT cryocooler built by Rutherford Appleton Laboratory for the Planck mission and predicted cooling power of a Lockheed-Martin Space Systems Co. [31] pulse tube cryocooler based on a fixed 200 W input power. For the latter, the number of stages is increased as shown as the operating temperature is reduced.

The result is that the cryocooler must tolerate rapid transitions between high and low heat flux, with the high value at or close to the physical limit based on design and input power constraints. In the transition to low heat flux, the drop in internal pressure creates the possibility that the compressor can be overdriven unless the input power is reduced. If such control is not built into the control electronics, it may then be necessary to restrict the ADR to a lower peak heat flux such that the cryocooler can safely operate as the heat flux transitions between extremes.

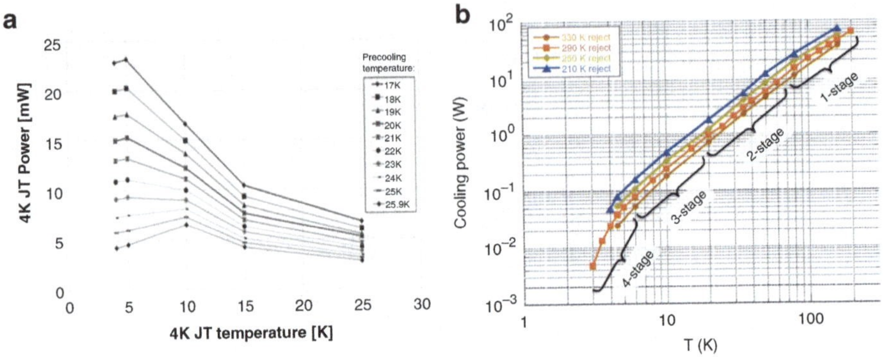

Fig. 5.5 Measured cooling power versus temperature for (**a**) the 4 K Planck JT cryocooler and (**b**) predicted performance for the Lockheed-Martin pulse tube cryocooler. (Credit: Jean-Loup Puget/Planck Collaboration and Lockheed-Martin Space Systems Co.)

Since the inevitable result is a reduction in cooling power and performance, other strategies have been investigated to smooth out the heat flow to the cryocooler. The first involves adding a "thermal ballast" at the interface between the ADR and cryocooler whose large effective heat capacity acts as a thermal storage unit, absorbing or giving off heat as the heat flux rises above or falls below a target value. One option for a passive thermal ballast consists of a sealed container of helium gas which can store and release heat through evaporation and condensation. The onset temperature (below which heat capacity is enhanced by condensation and evaporation) and the total heat capacity can be adjusted by the volume and room temperature charge pressure.

For continuous ADR systems, an alternative is to employ two stages in parallel [21] in which one is actively cooling the load while the other is rejecting heat. This implementation still produces a fluctuating heat load to the cryocooler, as continuous cooling requires periods when both the stages are at low temperature (and therefore not rejecting heat to the cryocooler), so that a handoff between them can occur. However, if a third stage is used (stage "c" in Fig. 5.6) to stabilize the temperature at the load, the two parallel stages can be operated, so that one of them is always rejecting heat. By maximizing the heat rejection rate (at a constant high value), the ADR can achieve the highest possible cooling power for a given cryocooler input power.

This strategy has been adopted for the Primordial Inflation Explorer (PIXIE) instrument, which is currently being proposed to NASA to measure the polarization of the cosmic microwave background, whose telescope must be continuously cooled to ~2.72 K with a constant heat load of 9.0 mW by 3-stage continuous ADR rejecting heat to a 4.5 K JT cryocooler. A schematic of the ADR is given in Fig. 5.6, which also includes a 3-stage continuous ADR for cooling the detectors to 100 mK. The latter is discussed in more detail in the next section.

Figure 5.7 shows a simulation of temperature and heat rejection when the ADR is operated in constant cooling mode and in constant heat rejection mode. The time

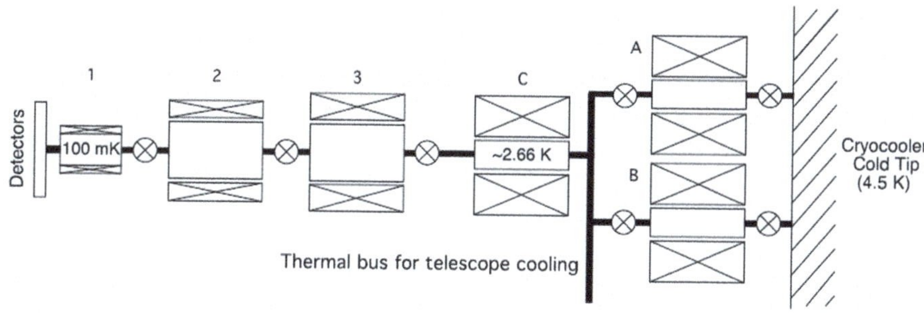

Fig. 5.6 Schematic of the PIXIE ADR, which contains two 3-stage ADRs that will operate continuously at approximately 2.66 K (to act as a heat sink for the telescope at 2.72 K) and 100 mK

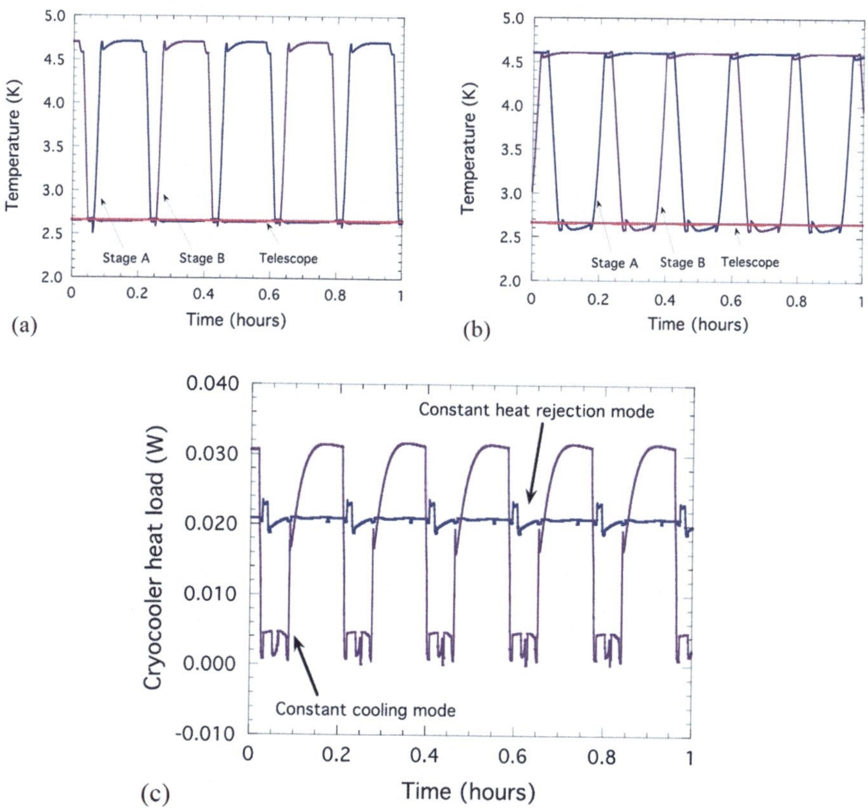

Fig. 5.7 Simulated temperature and heat rejection rates for continuous operation of a 3-stage continuous ADR. Temperatures of the two cycling stages and the stabilizing stage (connected to the telescope) are given for (**a**) constant cooling mode and (**b**) constant heat rejection mode. (**c**) A comparison of the instantaneous heat rejection rates for the two modes

average cryocooler loads are 19.8 and 20.8 mW, respectively, for continuous heat lift of 9 mW at 2.72 K. The difference reflects the lower efficiency of the constant heat rejection mode, as the ADR must absorb heat from the telescope more rapidly using a larger thermal gradient. Nevertheless, it can be seen that this mode results in a significantly lower peak cryocooler load of ~23 mW and peak to peak variations of ~3 mW, compared to a peak of ~32 mW. In principle this can translate to lower cryocooler input power and higher system efficiency.

5.3 Thermal and Mechanical Interacting to a Cryocooler

5.3.1 Thermal Links

The driving consideration for the connection between an ADR and its heat sink is its thermal conductance, as this will dictate the temperature at which the salt pills in the warmest stages must be regulated in order to generate the required heat flow. As previously noted, the latter will generally be in the range of 20–30 mW, depending on the capability of the cryocooler and the fraction of that capacity used for other heat loads. Excessive temperature differences between the salt pill and heat sink during magnetization decrease the efficiency of the ADR and can lead to other potential problems. One of the most serious of these arises from the fact that the magnet and salt pill typically share the thermal link to the heat sink, and a poor link that requires elevating the salt pill temperature will also elevate the temperature of the magnet. Especially for systems operating close to the limits of NbTi, the latter may drive the magnet beyond to its critical field and induce a quench.

For Astro-H, whose 3-stage magnet was designed to generate 3 T with a 2 A current, the maximum operating temperature was determined to be 5.2 K, requiring—for the sake of margin—that the maximum salt pill temperature during magnetization be $\leq$5 K. With the cryocooler required to operate at $\leq$4.5 K with a target heat flow from the ADR of 20 mW, the aggregate thermal conductance between the salt pill and cryocooler was required to be at least 40 mW/K. This conductance applies to the complete thermal path between the salt pill's thermal bus and the cryocooler, which in this case included a gas-gap heat switch, three separate thermal straps (two on the US side of the interface and one on the Japanese side), and a number of bolted joints. The thermal conductance of the heat switch was measured at 4.5 K to be approximately 200 mW/K, the salt pill's thermal bus and strap connecting it to the heat switch contributed 150 mW/K, and the strap between the heat switch and the US/Japanese interface was assumed to be 200 mW/K at the time of setting requirements, though in reality it exceeded 600 mW/K. The conductance of the remaining strap between the US/Japanese interface and the cryocooler was required to be at least 120 mW/K.

Flexible thermal straps help alleviate problems related to thermal contraction, and they provide compliance at the interface to facilitate the integration with the cryocooler and minimize mechanical coupling to any exported vibration from the cold head or compressors. The approach used for Astro-H [31] involved diffusion bonding eight 0.12-mm-thick high-purity copper foils (99.999%; <0.5 ppm Fe) to a 1-mm-thick sheet which had already been machined to its final shape for a bolted joint. The foils and sheet are pre-bent into their flight configuration and arranged with about 1 cm overlap between CRES 304 blocks that, when screwed together, compress the joint with 3 MPa pressure. The diffusion bonding is accomplished with a 2-h hold in vacuum at 800 °C followed by a 4-h cooldown to 250 °C (subsequent cooling and venting rates did not need to be controlled). All straps were then gold-plated.

The completed ADR and detector assembly, ready for integration into the Astro-H dewar, is shown in Fig. 5.8. In the final configuration, the connection between the flexible portion of the JT strap and the cryocooler was made with ultra-high-purity aluminum (>99.9999%) produced by Sumitomo Chemical Co. [32]. The flight assembly did not include the level of instrumentation needed to measure the thermal conductance of the entire link, though it could be confirmed that the conductance between the third stage salt pill and the U.S./Japanese interface was 75 mW/K. Provided that the interface is kept below 4.5 K during peak heat flow,

Fig. 5.8 The Astro-H ADR and detector assembly, suspended from its lifting sling, before shipment to Japan for integration into the Soft X-ray Spectrometer dewar. The flexible thermal strap to the 4.5 K JT cryocooler is in the foreground

regulating the salt pill at 4.9 K during magnetization yielded the desired transfer rate of 30 mW.

The Astro-H ADR and detector assembly is suspended from its lifting sling, before shipment to Japan for integration into the Soft X-ray Spectrometer dewar. The flexible thermal strap to the 4.5 K JT cryocooler is in the foreground.

5.3.2 *Optimization of Magnetic Current Leads*

Compared to most other cryogenic components, the magnets for ADR stages draw relatively high current. Even the very-high-field-to-current magnets used in space ADRs require 2–4 A to reach full field. In resistive wire, this can lead to large joule (I^2R) heating. The heat generated in the leads and the thermal conduction down the leads from higher temperature must be minimized. Fortunately, at low temperature, standard NbTi wire can be used with negligible heating in the leads and at properly made joints. At higher temperatures, commercial high temperature superconductor (HTS) leads can be used with negligible dissipation providing that the joints have very low resistance. Sub-μΩ resistances can typically be achieved at each joint. These HTS leads can be purchased with an Ag-Au coating which has relatively low thermal conductance. In total, a typical pair of leads can have a few microwatts of thermal conductance and a few microwatts of peak dissipation down to 4.5 K. Typical specifications for the smallest commercial size HTS tape are 80 A maximum current at 77 K, so for typical ADR magnets, there are no concerns with exceeding the HTS critical current.

HTS leads can provide near-zero dissipation at temperatures up to about 80 K. Above this temperature, normal resistive wire must be used. A balance between joule heating and passive conduction (k) must be struck to minimize the thermal impact to the system from these resistive leads. For space flight cooling systems, the upper bound for use of HTS leads may actually be lower than 80 K, depending on the cooling stage available. For instance, in ASTRO-H the cryocoolers provided cooling at ~100 and ~30 K. Because the HTS material could not be guaranteed to work at temperatures above 80 K, the HTS was transitioned to resistive leads at the lower, 30 K stage. The lower the temperature, the lower the cooling capacity available, so optimization of the leads for the lowest overall heat is required. The heat generated by joule heating and transported by thermal conduction to the cold heat sink is given by

$$Q_{\text{dot}} = < I^2R > + \int k(T)dT \tag{5.2}$$

where we have included the time-varying nature of the joule heating and the subsequent induced temperature profile by taking a time average value indicated by "<>"symbols. We also assume a wire with a nearly constant resistance over temperature, such as phosphor-bronze (Ph-Br) that allows the optimization of the wire size to be independent of the temperature in the middle of the wire to first order. That is, if the resistivity of the wire changes significantly with temperature, then the diameter of the wire would also have to vary along the length to minimize the overall heat transported to the cold heat sink. Resistance, R, and thermal conductance, k, for metals are connected by the Wiedemann–Franz relation [32, 33]:

$$R = (L_0 T)/k, \tag{5.3}$$

where R is in units of ohms, k in W/Kelvin, T in Kelvin, and L_0 is the Lorenz constant, ~2.44 × $10^8 V^2/K^2$. For R approximately independent of temperature, Eq. (5.2) then becomes:

$$Q_{\text{dot}} = < I^2 R > + L_0 T_{\text{avg}} \partial T/(2R) \tag{5.4}$$

We can solve for the resistance that produces the minimum heat by taking the derivative of [Eq. 5.4] and setting it to zero. This gives:

$$R^2 = L_0 T_{\text{avg}} \partial T / I^2 \tag{5.5}$$

For example, take the ADR cycle as shown in Fig. 5.9. The time-averaged square of the current, $<I^2> = 2A^2$ or $I = \sqrt{2}$A and use 30 K as the cold heat sink and 100 K as the warm heat sink. $T_{\text{avg}} = 65$ K and $T = 70$ K. Using Eq. (5.5), we get that the optimum lead resistance is 7.5 mΩ, and the minimum average heat flow along this wire to the cold heat sink is 22 mW. The return lead generates a similar amount of heat to the cold stage for a total of 44 mW for a single ADR stage. Such an amount of heat can be readily handled by a typical flight cryocooler 30 K stage.

Radiation from the wires may be a potential consideration in the wire selection as well, but for a relatively high duty cycle, as in this example, the radiation from wire heating is negligible as seen in Fig. 5.10.

For low duty cycle operation, practical considerations, such as overheating of the resistive leads or temperature instability of the cold sink of the resistive leads,

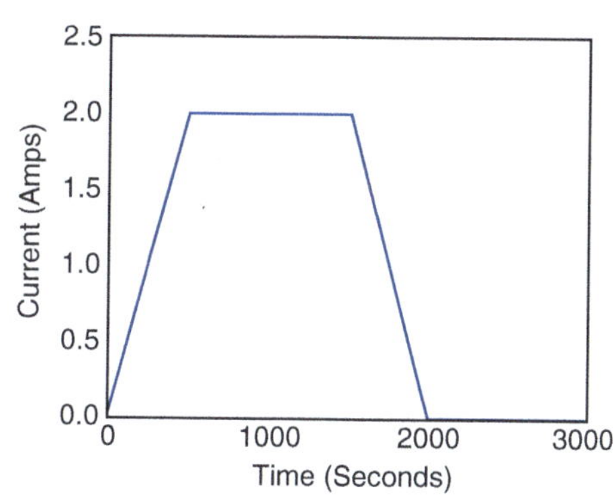

Fig. 5.9 A simplified example of current vs. time for an ADR cycle

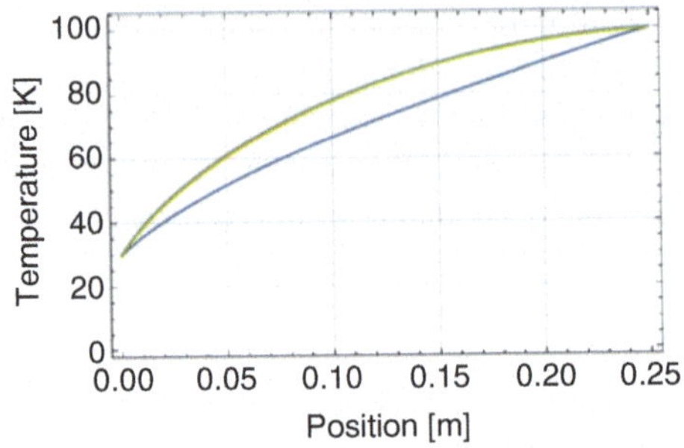

Fig. 5.10 Calculation of the temperature vs. position for a 250-mm-long optimized resistive wire between 100 and 30 K carrying a constant current of $\sqrt{2}$ A. The lower curve shows the value with no current flowing. The upper curve includes the joule heating for an average current of $\sqrt{2}$ A. A third curve that adds in radiation from the wire is indistinguishable from the upper curve

may limit the optimization and lead to smaller than optimum wire resistance. For instance, studies have been performed for Ph-Br wire used for the James Webb Space Telescope instruments demonstrating a safe upper limit on current for a given diameter wire suspended in a vacuum and covered with multi-layer insulation. On Astro-H, the low duty cycle Stage 1-resistive leads heated the 30 K heat sink resulting in a several Kelvin fluctuation with a time constant of several hours.

5.3.3 Thermal and Mechanical Considerations for Multistage ADRs

The discussion in Sect. 5.2.4 demonstrated how multistaging, and especially continuous operation, can provide a better match at the thermal interface between an ADR and spaceflight cryocoolers. Multistaging carries with it other possibilities for improved performance, but also introduces complexities for integrating with the cryocooler and the warmer cryogenic system. These relate to the thermal and mechanical configuration of the stages and the need for electrical leads to deliver the current required by each stage.

In the vast majority of ADR implementations, the stages are configured with their magnets thermally attached to the heat sink, which directly absorbs magnetic hysteresis heat and heat switch dissipation. For ADRs consisting of more than one stage, the salt pills can also be linked directly to the sink through their heat switches or linked in series. Regardless of the latter, the structural support of the salt pills terminates at the heat sink, requiring the suspension system to span from the heat sink temperature to the operating points of the salt pills. For laboratory ADRs using commercial cryocoolers that typically produce a base temperature nearly 3 K, this can be easily managed. However, for space cryocoolers, the more typical operating temperature is 4.5 K, from which conduction in the suspension system can be excessive.

Of course it is possible to integrate a heat intercept into the suspension systems that are cooled by warmer stages, thereby reducing the heat load to the colder stages. But in multistage systems, it can be impractical to integrate multiple intercepts not only in each ADR stage but in instrument components being cooled, especially in the compact geometries required in space instruments, and instead, it may be advantageous to operate one or more stages from a lower temperature produced by other ADR stages. The practicality of doing so rests largely on the temperature range of the ADR but may also be necessary in situations where large heat capacities are being cooled, as the arrangement can significantly reduce the temperature excursions of each stage and the amount of cooling capacity used during demagnetization.

The penalty of thermally anchoring an ADR stage to a temperature lower than that of the heat sink—i.e., to one produced by other ADR stages—is that magnetic hysteresis and heat switch dissipation increase the heat lift requirements for those upper stages. The benefit is a significant reduction in the heat loads to the colder stages and in the magnetic fields required, thereby requiring smaller salt pills and magnets and magnetic shields, which in turn generate less hysteresis heat.

When an ADR is operated in single-shot mode, thermally anchoring any stage at a temperature below the sink would have its magnet fluctuate in temperature during (de)magnetization. Especially when ferromagnetic shielding is used, the heat capacity that must be cooled would likely use an excessive fraction of the cooling capacity of the ADR during demagnetization. Consequently, the situations where it makes most sense to anchor stages at lower temperature than the sink are when continuous cooling is employed. The concept for PIXIE discussed above (Fig. 5.6) uses three stages to continuously cool the telescope to ~2.72 K. An additional three stages are used to continuously cool the detectors to 100 mK using the telescope as its heat sink.

Supporting the detector ADR from a ~2.7 K, instead of 4.5 K, platform reduces the heat load to the warmest stage by at least a factor of 3, as the thermal conductance of materials typically used in suspending salt pills—e.g., Kevlar fiber [34], Vespel SP-1 or SP-22 [35], carbon fiber composites—has a temperature dependence between T^{-2} and T^{-3}. Other specialty materials, such as Ti15333 [36], have superconducting properties that can yield exponential temperature dependences, particularly at very low temperature. Based on the examples in Table 5.2, for a stage operating between ~0.5 and 2.7 K, we expect at least 70% of the heat rejected from the warm stage to originate in the salt pill, so that the total heat rejected will be approximately 1.4 times the salt pill heat. By reducing the parasitic heat load by more than a factor of 3, the total heat load from the warm stage of the detector ADR is reduced by at least a factor of 2.

The advantages of connecting to a lower base temperature extends to the colder ADR stages, as well as any instrument components, such as detector assemblies and amplifiers. In addition, these components can make use of the warmer stage as a heat intercept to further reduce parasitic heat loads and consequently the mass and heat output of the ADR. It should be noted in the example of PIXIE that the choice of intermediate temperature is driven by the need to cool the telescope to 2.72 K in order to match the temperature of the cosmic microwave background. In fact, the advantage to system mass and efficiency continues to improve as the intermediate temperature is reduced. The point of diminishing returns is reached when stages must be added to the upper ADR assembly in order to span a larger temperature range. Guidelines for determining the number of stages and their optimization can be found in [37]. From a 4.5 K heat sink, the optimal intermediate temperature is in the range of 1–1.5 K.

5.3.4 Future ADR Developments and the Impacts for Space Cryocoolers

The operating efficiency, in terms of input power per watt of cooling power, of spaceflight cryocoolers [38]. At low temperature, the specific power has a $1/T^2$ dependence. In comparison, ADRs achieve near Carnot efficiency, yielding a specific power with a linear temperature dependence. This suggests that one direction for future development is to raise the operating range of ADR systems and couple them to cryocoolers operating at higher temperature. One example implementation would use a 2-stage continuous ADR operating at 4–5 K coupled to a cryocooler at 10 K. Any existing systems operating at 4–5 K could be directly coupled to this hybrid cooler and achieve as much as a factor of 2 improvement in overall efficiency. Operation at even higher temperature may be feasible, but as noted previously, the rising non-magnetic portion of the heat capacity of the refrigerant would eventually require a significant number of stages in series to produce the required temperature excursion.

The main challenge is the development of magnets capable of producing the fields needed (3–4 T) at temperatures above 10 K. The development of advanced refrigerant materials tailored to produce larger entropy change per field in the 5–10 K range, as well as improvements in heat switches and suspension systems, would also be beneficial. For cryocoolers, the development would expand the range of operating temperature that can be used with ADR systems and reduce the cooling requirements—potentially enabling simpler, lower power cryocoolers to be selected for space missions, but also potentially allowing the performance margin that would be gained by higher temperature operation to offset the "cost" of implementing load-leveling capabilities within the cryocooler.

References

1. Moseley SH, Mather JC, McCammon D (1984) Thermal detectors as X-ray spectrometers. J Appl Phys 56(1984):1257–1262
2. Irwin KD, Hilton GC, Wollman DA, Martinis JM (1996) X-ray detection using a superconducting transition-edge sensor microcalorimeter with electrothermal feedback. Appl Phys Lett 69:1945

3. Smith SJ et al (2016) Transition-edge sensor pixel parameter design of the microcalorimeter array for the X-ray Integral Field Unit on Athena [Space Telescopes and Instrumentation 2016: ultraviolet to gamma ray]. In: den Herder JW, Takahashi T, Bautz M (eds) Proceedings of SPIE, vol 9905, 99052H
4. Gottardi L, Akamatsu H, Bruijn M, den Hartog R, den Herder J-W, Jack- son B, Kiviranta M, van der Kuur J, van Weers H (2016) Development of the super-conducting detectors and read-out for the X-IFU instrument on board of the X-ray observatory Athena. Nucl Instrum Methods Phys Res Sect A 824(11):622–625
5. Doriese WB et al (2016) Developments in time-division multiplexing of X-ray transition-edge sensors. J Low Temp Phys 184:389–395
6. Bender AN et al (2014) Digital frequency domain multiplexing readout electronics for the next generation of millimeter telescopes. In: Holland WS, Zmuidzinas J (eds) Millimeter, submillimeter, and far-infrared detectors and instrumentation for astronomy VII. Proceeding of SPIE 9153, 91531A
7. Dobbs MA et al (2012) Frequency multiplexed superconducting quantum interference device readout of large bolometer arrays for cosmic microwave background measurements. Rev Sci Instrum 83(7):073113
8. Irwin KD, Cho HM, Doriese WB et al (2012) Advanced code-division multiplexers for superconducting detector arrays. J Low Temp Phys 167:588–594
9. Barcons X, Nandra K, Barret D, den Herder J-W, Fabian AC, Piro L, Watson MG, the Athena Team (2015) Athena: the X-ray observatory to study the hot and energetic universe. J Phys Conf Ser 610:012008
10. Harvey PM (1979) A far-infrared photometer for the Kuiper Airborne Observatory. Astron Soc Pac Publ 91:143–148
11. Pilbratt GL, Riedinger JR, Passvogel T, Crone G, Doyle D, Gageur U, Heras AM, Jewell C, Metcalfe L, Ott S, Schmidt M (2010) Herschel Space Observatory: an ESA facility for far-infrared and submillimetre astronomy. Astron Astrophys 518:L1
12. Planck collaboration (2011) Planck early results. II. The thermal performance of Planck. Astron Astrophys 536:A19
13. Duband L, Clerc L, Ercolani E, Guillemet L, Vallcorba R (2008) Herschel flight models sorption coolers. Cryogenics 48:95
14. Triqueneaux S, Sentis L, Camus P, Benoit A, Guyot G (2006) Design and performance of the dilution cooler system for the Planck mission. Cryogenics 46:288
15. Kelley RL et al (2007) The Suzaku high resolution X-Ray spectrometer. Publ Astron Soc Japan 59:S77–S112
16. Serlemitsos AT, SanSebastian M, Kunes E (1992) Design of a spaceworthy adiabatic demagnetization refrigerator. Cryogenics 32:117
17. Sugita H, Sato Y, Nakagawa T, Murakami H, Kaneda H, Enya K, Murakami M, Tsunematsu S, Hirabayashi M (2008) Development of mechanical cryocoolers for the Japanese IR space telescope SPICA. Cryogenics 48:258–266
18. Shirron PJ, Canavan ER, DiPirro MJ, Tuttle JG, Yeager CJ (2000) A multi-stage continuous-duty adiabatic demagnetization refrigerator. Adv Cryo Eng 45:1629–1638
19. Lounasmaa OV (1974) Experimental principles and methods below 1 K. Academic, London
20. Pobell F (2007) Matter and methods at low temperatures, 3rd edn. Springer, New York, pp 203–213
21. Shirron P (2014) Applications of the magnetocaloric effect in single-stage, multi-stage and continuous adiabatic demagnetization refrigerators. Cryogenics 62:130–139
22. Tuttle JG, Pourrahimi S, Canavan ER, DiPirro MJ, Shirron PJ (2006) A lightweight low-current 10 K magnet for space-flight ADRs. Cryogenics 46:196–200
23. Burdyny T, Arnold DS, Rowe A (2014) AMR thermodynamics: semi-analytic modeling. Cryogenics 62:177–184
24. Hagmann C, Richards PL (1994) Two-stage magnetic refrigerator for astronomical applications with reservoir temperatures above 4K. Cryogenics 34:221

25. Shirron PJ, Kimball MO, Fixsen DJ, Kogut AJ, Li X, DiPirro MJ (2012) Design of the PIXIE adiabatic demagnetization refrigerators. Cryogenics 52:140–144
26. Shirron PJ, Kimball MO, James BL, Wegel DC, Martinez RM, Faulkner RL, Neubauer L, Sansebastian M (2012) Design and predicted performance of the 3-stage ADR for the soft-X-ray spectrometer instrument on Astro-H. Cryogenics 52:165–171
27. Shirron PJ, Kimball MO, James BL, Muench T, DiPirro MJ, Let- mate RV, Sampson MA, Bialas TG, Sneiderman GA, Porter FS, Kelley RL (2016) Operating modes and cooling capabilities of the 3-stage ADR developed for the soft X-ray spectrometer instrument on Astro-H. Cryogenics 74:2–9
28. James BL, Martinez RM, Shirron PJ, Tuttle JG, Francis JJ, SanSebastian M, Wegel DC, Galassi NM, McGuinness DS, Puckett D, Flom Y (2012) Mechanical design of a 3-stage ADR for the Astro-H mission. Cryogenics 52:172–177
29. Wilson M (1983) Superconducting magnets. Oxford University Press, New York, p 162
30. Wikus P, Canavan ER, Heine ST, Matsumoto K, Numazawa T (2014) Magnetocaloric materials and the optimization of cooling power density. Cryogenics 62:150–162
31. Olson JR, Roth E, Champagne P, Evtimov B, Nast TC (2008) High performance pulse tube cryocoolers. Adv Cryo Eng 53A:514–521
32. Kittel C (1971) Introduction to solid state physics, 4th edn. Wiley, New York, p 263
33. Tuttle JG, Canavan ER, DiPirro MJ (2010) Thermal and electrical conductivity measurements of CDA 510 phosphor bronze. Adv Cryo Eng 56A:55–62
34. Ventura G, Barucci M, Gottardi E, Peroni I (2000) Low temperature thermal conductivity of Kevlar. Cryogenics 40:489–491
35. Locatelli M, Arnaud D, Routin M (1976) Thermal conductivity of some insulating materials below 1 K. Cryogenics 16:374
36. Wikus P, Hertel SA, Leman SW, McCarthy KA, Ojeda SM, Figueroa-Feliciano E (2011) The electrical resistance and thermal conductivity of Ti 15V-3Cr-3Sn-3Al at cryogenic temperatures. Cryogenics 51:41–44
37. Shirron PJ (2014) Optimization strategies for single-stage, multi-stage and continuous ADRs. Cryogenics 62:140–149
38. Johnson D. Cryogenic technology for CMB-Pol: mechanical cryocoolers for the 4K to 200K temperature range, CMB Polarization Workshop, Boulder, CO, August 27, 2008. http://cmbpol.uchicago.edu/workshops/technology2008/depot/johnsoncryocoolers.pdf¿.

Chapter 6
Cryocoolers for Superconducting Generators

Wolfgang Stautner

Abstract Historically, the cooling of superconducting generators has been an immensely complex and expensive undertaking. The process requires large, high-power, stationary helium liquefaction plants, usually situated separately from a generator, using long transfer lines to supply a continuous liquid helium flow to the generator. The lack of a suitable, economically viable cryogenic infrastructure and the resulting technical challenges, e.g., the need for a rotary transfer coupling, proved to be stumbling blocks on the path to successful commercialization. As of today, and nearly 50 years later, initial generator cooldown time, coupled with mean time between failure service requirement and operational recooling after an outage are still the main cryogenic concerns. The following chapter touches briefly on the early beginnings of the process and outlines the technological efforts and progress made in this respect until now. It further illustrates how the exceptional modularity of today's cryocoolers, as manifested by their siteability, by the option of combining cryocoolers with different cooling capacities depending on their cryogenic system design and by their greatly improved, long maintenance intervals, enables the advancement of superconducting generator technology, accelerating its market readiness.

6.1 Introduction

GE Global Research can look back to a long history in the development of superconducting technology that brought together some of the most important scientists in this field. For example, Ivar Giaever who received the Nobel Prize in 1973 for his work on electron tunneling in superconductors, or Charles Bean who developed the Bean model for current propagation in superconductors in 1964, which formed the basis for all current understanding. Not to mention all other

W. Stautner (✉)
GE Research, Biology and Applied Physics, Cryogenics & Superconducting Magnet Applications, One Research Circle, Niskayuna, NY, USA
e-mail: stautner@ge.com

M. D. Atrey (ed.), *Cryocoolers*, International Cryogenics Monograph Series,
https://doi.org/10.1007/978-3-030-11307-0_6

scientists occupied with groundbreaking work on NbTi and Nb_3Sn, for example, deSorbo and Healy who reported first on the intermediate state of some Type I and Type II superconductors as early as 1964 [1] using improved magneto-optic techniques for rendering visible this intermediate state and analyzing lead, tantalum, niobium in pure and alloyed form. They also reported on the influence of impurities and solutions on the high-current carrying capacity Nb_3Sn superconductor in 1964 [2]. Consequently, GE leveraged the spin-off of IGC to produce Nb_3Sn tapes in 1970.

As for applications and components, Buchhold first suggested the use of superconductive bearings for three phase motors in 1965 [3] and GE built the first SC flux pump in August 1967 [4]. Buchhold also first suggested cryogenic design approaches for superconductive helium pumps [5].

The development of superconducting rotating electrical machines began in this exciting period of time. The understanding of the behavior of multi-filamentary NbTi began to grow and the first wires with a reasonable length started to become available at the end of the 1960s. From that time onwards until the 1990s, several AC synchronous machines were successfully tested at GE with respect to their performance for a wide power range.

The difficulty here was to maintain an operating temperature of preferably 4 K or lower for the LTS type conductor. Thus, one had either the choice to efficiently cool the field winding with forced-flow cooling techniques or use rotating thermosiphons with enhanced heat transfer to the winding supported by the centrifugal forces or by simply immersing the field winding in a helium bath as with today's MRI magnet systems. The first two options mentioned were considered feasible for rotating field windings, which were the choice for operating an SC generator until the 1990s.

Since suitable cryocoolers did not exist then, those early cooling technologies consisted of bulky and heavy refrigerating supplies for 4–1.8 K of continuous helium mass flow through pipelines to the cooling object. Because of its size requirements, the complete refrigerating system with its components had to be housed in a separate building. Economic considerations, amongst others that had to include the cooling technology effort, made these machines unattractive at that time.

With hindsight, it seems that the cryogenically most challenging and complex solution was chosen for the SC generator early on and was therefore bound to fail. The huge number of patent applications in this field show the frantic effort made to find workable cryogenic solutions.

After a long pause, renewed efforts were made to get rotating machinery into the utility market and into the grid once the new type 1G and 2G superconductors became available in longer lengths and when operating temperatures jumped from 20–70 K.

In parallel and right on time, cooling technology in the form of different coolers of the GM, PTR, and Stirling type improved significantly, thereby opening up new technical solutions.

Even better, rotor and stator dynamics were revisited and a more open approach chosen based on the newer computational tools that were now available, combined with new materials and assembly and production techniques.

6.2 Benefits of High Power Density SC Generator Technology

So what are the current main drivers for leveraging superconducting vs conventional field proven technology?

With respect to the

Generator:

- High efficiency under full or partial load conditions
- Supplies full reactive power rating
- The overall SC generator weight is drastically reduced and packaging most compact
- Lifetime performance is far higher compared with conventional systems owing to the absence of field winding current changes when exposed to environment (in most cases the superconducting winding is housed in a cryostat and a vacuum chamber and therefore thermally protected from environment)
- Potential of retrofitting an SC component into a conventional stator
- Can increase the air gap (important for some applications)

and for the

Grid Interconnection:

- Short term overload capability can be very high, combined with
- an excellent transient stability that fits right into the grid

As for the stator design, the main choices are either the use of an SC rotor with a toothed conventional stator for the flux guidance or no magnetic iron teeth with air gap winding.

Last but not least, superconducting technology illuminates the future of high-powered devices and structures that greatly surpass today's conventional engineering practices.

6.3 Early High Power SC Generators: Examples

LTS-based SC generator programs started in the USA, France, Japan, and Germany early on in the late 1960s. Figure 6.1a–h show superconducting generators developed and built by GE in the 1970s. The assembly of one of the very first 20 MW LTS-based SC generators operating at a liquid helium temperature of 4 K is shown in (a) and (b). As mentioned, GE analyzed the SC material properties attractive for SC electric machinery very early on in 1965. With the arrival of Nb_3Sn, a 20 MW NbTi/Nb_3Sn hybrid generator operating at 4.5 K and 8 K, respectively, was built for military applications (c). Those efforts were followed by the design of a DC drum disk-type homopolar motor for propulsion systems running at 1200 rpm and 3000 hp. Using an HTS conductor, 1.5 MW generators (demo systems) were built and successfully tested, shown in (d) and (e).

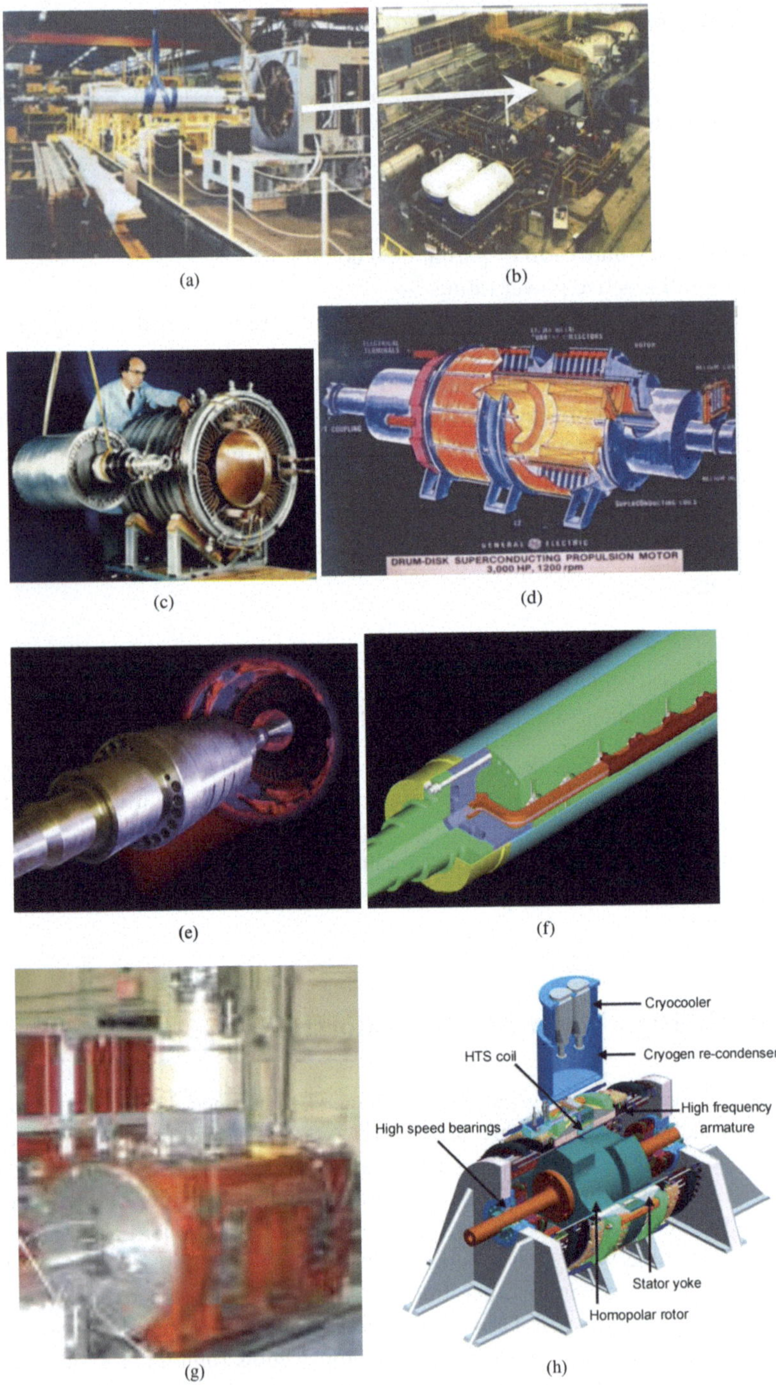

Fig. 6.1 (**a**–**h**) GE generator technology from LTS to HTS

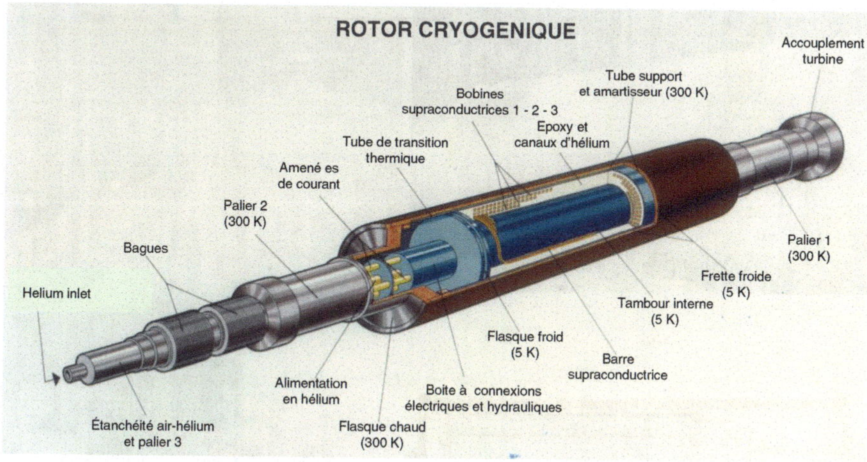

Fig. 6.2 Alsthom rotor design of 1972 [8]

Figure 6.1f also shows the 3D cutaway view of the HTS field rotor for a design study of a 100 MW HTS generator for utility applications as part of DOE program DE-FC36-02GO11100 [6].
(Color designation: field coil in red, helium duct marked in pink, magnetic pole in green and non-magnetic spacer in dark blue, U-Channel shown in brick red, and electromagnetic shield in light blue.)

In 2005, GE successfully ran a 5 MW HTS-HIA generator at 35,000 rpm at an operating temperature of 30 K and, following this, a 4 MW demonstrator running at 10,500 rpm was built and tested (see also Figs. 6.1g, h and 6.18) [7].

In 1972 the so-called cryogenerator program was kicked off by EDF and Alsthom (now GE Power) with the aim of manufacturing and testing a 300 MVA rotor based on NbTi/copper winding (Fig. 6.2).

The field at the winding was 4 T with an operating temperature of 5 K. Running at a high current of 5000 A and at 3000 rpm, the rotor ran successfully in the factory for its scheduled test period of 3 months (Fig. 6.3).

Liquid helium entry is shown in Figs. 6.2 and 6.3 at the generator tip (shown in yellow). Figure 6.2 shows the thermal shields around the helium inlet tube, expansion bellows and superconducting windings in the center. At the top right the warm end coupling to the turbine is facilitated.

6.4 Basics of SC Generator Design

Figure 6.4 shows the most basic design as well as several components of an early NbTi-based superconducting generator with its operating temperature at 4 K.

Fig. 6.3 Final assembly and inspection of the Alsthom rotor (weight 15,000 kg) [8]

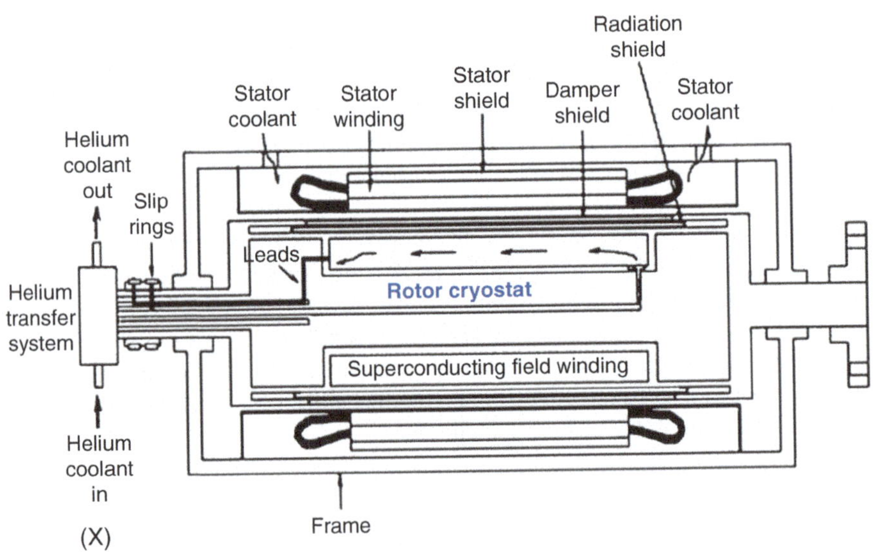

Fig. 6.4 LTS SC generator basic components with rotor cryostat [9]

Rotor and stator are integrated into a cylindrical, tubular shaped housing, called frame. The outer stationary, non-superconducting stator winding is fitted into a separate chamber within the frame and fitted with an outer stator shield and a damper shield on the inner radius to screen against field penetration from the inner superconducting field winding. Since the resistive stator winding is experiencing high temperatures during operation, the stator receives an air cooled flow via entry and exit slots.

The difficulty of a rotating inner winding starts with the need for a vacuum tight feedthrough (shown on the left) which calls for special sealing techniques of ferrofluid type. Moreover, the leads for the slip ring that grab the power from the superconducting magnet have to be fed through the same tubing as well.

Looking now at the helium supply end of the generator on the left in Fig. 6.4 transfer of liquid helium starts at point (X) and is fed into the far end of the field winding housing where the liquid boils off within the winding chamber and returns to the helium transfer system.

The superconducting field winding within the rotor cryostat is shielded against the influx of thermal radiation from room temperature and from the heat influx from the stator winding.

The engineering effort required for this design should not be underestimated and it needs a good understanding of how the rotor cryostat has to be designed to minimize steady state and transient heat loads and to understand how helium flow and heat exchange with the SC winding work together and can be optimized.

Many different cooling options have been derived from this design in the past.

6.5 Early Generator Cryogenics

Above all, a good understanding of the specifics of rotating heat transfer of the coolant and associated friction losses is mandatory when designing rotating machinery for highest efficiency with respect to forced-flow cooling of the superconducting field winding.

Andvig in 1956 [10] and Morris [11] in 1964 already showed that thermosiphon technology could be the simplest and most practical approach. One of the first rotational thermosiphon test setups that could be used to gain insight into the cooling of rotational field windings (although not for cryogenic use) is given in Fig. 6.5.

In this figure, the legs of section A-B-C-D of the rectangular frame could be filled with the test fluid (e.g., air), whereas the inner coiled tube stretching from D to C (length of 304 mm) is cooled by a secondary fluid, e.g., water. With hindsight, not surprisingly, despite its simplicity, it already contained some of the main design problems SC technology and Cryogenics have to solve when opting for these design features, namely good heat transfer efficiency from A-D or B-C to the central axis, and from the mechanical and electrical perspective, the design of bearings, rotary seals, the overall frame design and slip rings. The latter were used to transfer current into the heater assembly extending from section C and D within the tube. The rotor could rotate at speeds of up to 300 turns/min.

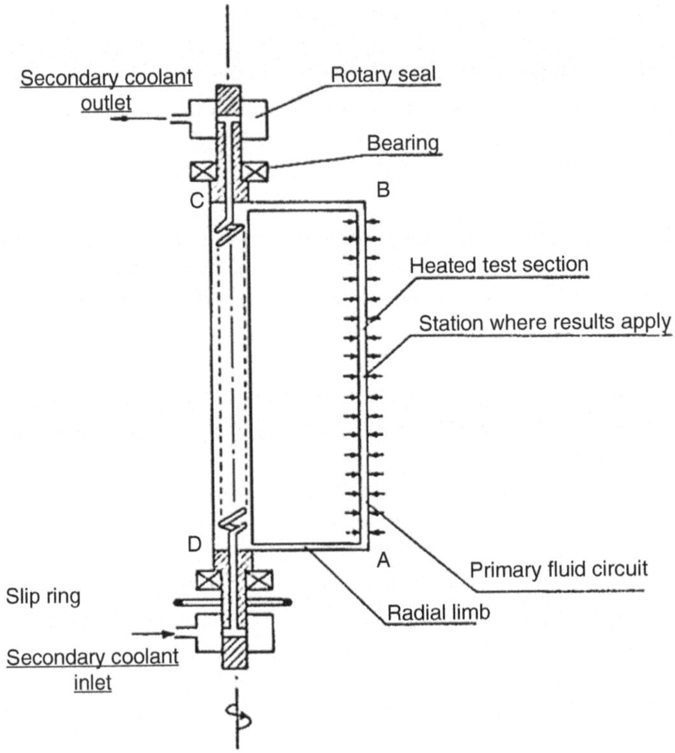

Fig. 6.5 Early thermosiphon gas flow test setup for rotor cooling [11]

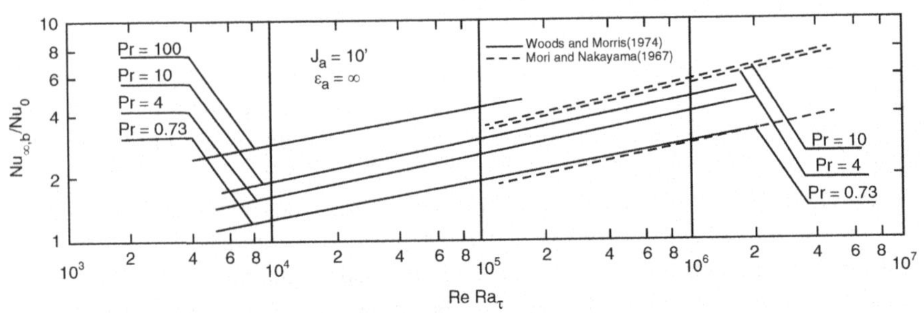

Fig. 6.6 Heat transfer (Nub/Nuo) vs (ReRa) numbers for different Pr numbers [11]

See also Lock [12] for further details on the working principles of rotating thermosiphons.

Figure 6.6 shows some of the test results obtained with this device for different Prandtl numbers (e.g., air Pr = 0.73). Here the difference between wall temperature and bulk fluid temperature Nu_b vs the classic Nusselt solution Nu_o as suggested by Mori and Nakayama [13] is shown.

A later, further improved version of this test setup allowed the coolant being drawn into the legs by an end pipe leg section designed as a centrifugal pump, a precursor of what was later suggested by Hofmann for the use of cooling superconducting windings and cryogenic fluids [11].

Still without the comfort of buying in compact coolers off-the-shelf (see also Table 6.2), for superconducting generator technology to be successful, be it for ship propulsion or for harvesting wind energy, or turbomachinery applications, not only the superconducting magnet design but packaging and integrating the vast cryogenic infrastructure was considered a major task: specifically, the ability to service and maintain its many different cooling components.

Since the early superconducting designs started right from a rotating field winding, the cryogenic environment had to rotate with it as well. The challenge here was that rotating heat transfer as started early by Morris with air and at room temperature was practically unknown when using 2-phase helium even though the Prandtl numbers of air and helium are similar.

Nevertheless, it was Hofmann et al. [14] and Schnapper [15] who developed rotor cryostat test setups that allowed convenient self-filling from a liquid helium Dewar. Those experiments gave a first glimpse in to how the cooling of rotational structures might project (extrapolate) for large generators (see Fig. 6.7 [15]). This simplified rotational thermosiphon setup shown in Fig. 6.7 allows to assess friction flow and heat transfer within the legs. With the results, it is then possible to determine the Rayleigh number.

Table 6.1 shows some typical design parameters for the loop shown in Fig. 6.7. Note that two types of legs are possible, either in parallel where legs have the same

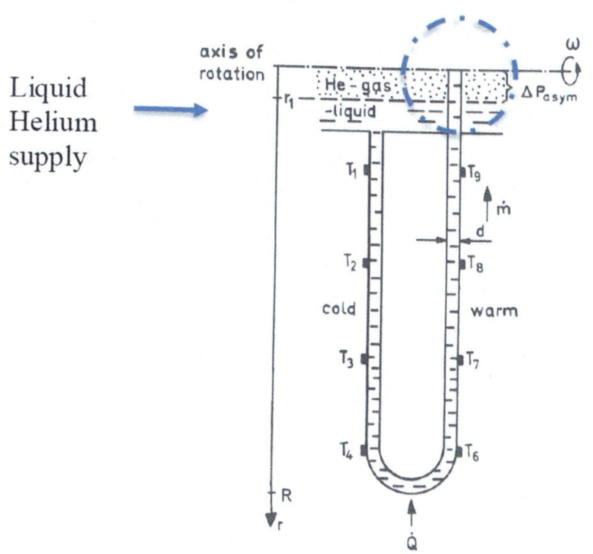

Fig. 6.7 Thermosiphon loop legs of a rotor cryostat [15]

Table 6.1 Test variables of an early rotor cryostat [15]

Variable	Values
Channel tube diameter d (mm)	1.35; 3.0; 6.0; 15.0
Rotational speed f (Hz)	16.6; 33.3; 50 and intermediate values
Heating power Q (W)	0–3
Distance of sensors T from rotational axis (m)	0.07–0.33

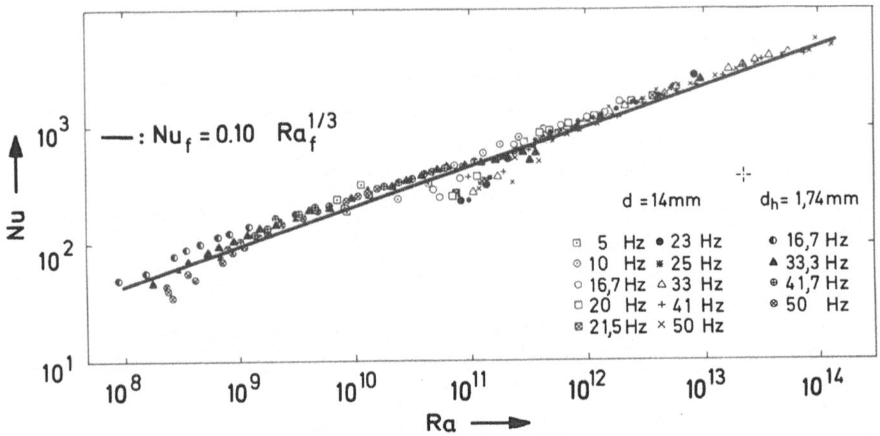

Fig. 6.8 Cryogenic rotating heat transfer (in memoriam of C. Schnapper) [15]

lengths or asymmetric where one leg extends to the rotating center filled with helium gas (see dotted line).

A typical test result is given in Fig. 6.8. Thermosiphon loops are employed to efficiently transfer heat from the rotor periphery to the central axis of the rotor. The above curve in Fig. 6.8 shows the almost linear increase in heat transfer with increasing rotational speed for helium from 4.2 to 5.17 K.

Here, the Rayleigh number for rotational systems replaces the gravitational acceleration with the centripetal acceleration for the thermosiphon and is conveniently written as $\mathrm{Nu} = a\mathrm{Ra}^b$. Figure 6.9 shows an early self-filling rotor cryostat. The stationary tubing from the supply Dewar is shown inserted angled into the rotary cryostat wall.

The following inset in Fig. 6.10 shows a further advanced practical design realization of the aforementioned efforts and highlights its corresponding efficiency in the h/s diagram.

The coolant enters the rotor cryostat at nominal pressure of 1 bar and at 4.2 K. Within the rotor, and depending on the rotational speed, the pressure can rise up to 20 bar and temperatures above 5.5 K. Since this temperature normally is at the high end for a respective economical design, the boil off is routed through thermosiphon legs that basically remove the gas from the windings by performing like a centrifugal pump (flow path 7-8-9-10). Good designs are thus capable of

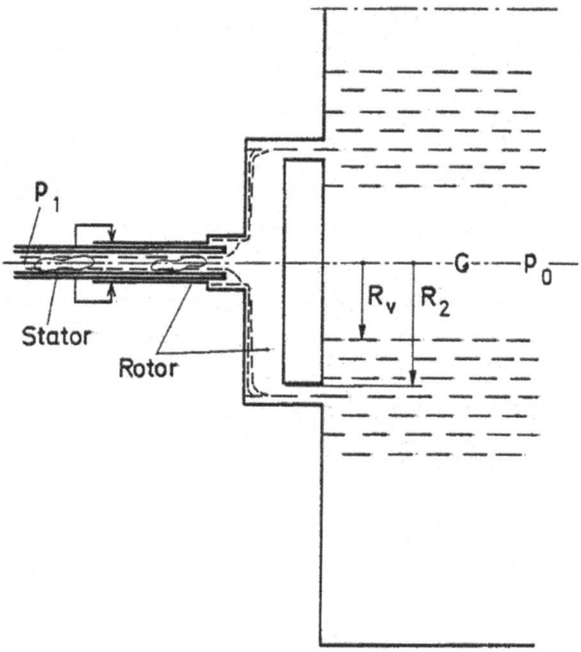

Fig. 6.9 Rotor cryostat (after Hofmann [14])

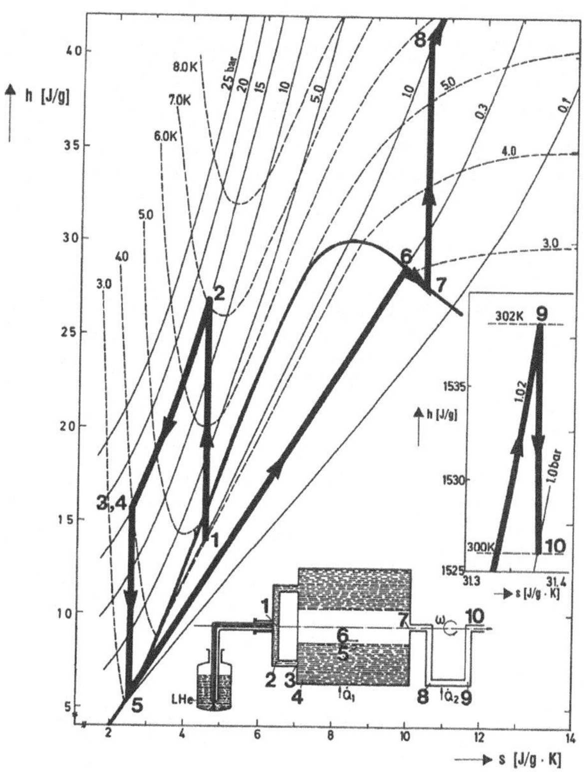

Fig. 6.10 Advanced cryogenic rotating heat transfer for reducing helium temperature in a rotor cryostat (courtesy of KIT)

achieving subatmospheric temperatures. Temperatures close to 3 K are possible which leads to a high enough margin with respect to the superconducting current sharing temperature [16, 17] for NbTi wires.

6.6 Cryocooler Rotor Cooling Technologies for SC Generators and Motors

Depending on the design environment and given constraints the strategy for the cooling of superconducting generators may mainly follow the design approaches as shown below:

For reasons of simplicity, we need to further refine "Remote cooling." Basically this means, superconducting generator and cold box/refrigerator unit are physically separated from each other by transfer lines for liquid or gaseous cryogens. The coolant medium is transferred to the field winding by means of impellers, fans, or pumps.

"Conduction cooling" means that the cryocooler cold stage is in direct physical contact with the cold mass. If the cryocooler is sleeved or any other conducting interface (e.g., heat pipe, conducting medium) is fitted in between both cooler and cold mass, this contact is called "indirect."

"Thermosiphon cooling," on the other hand, can be facilitated remotely as well as close to the field winding [18] and falls more into the indirect cooling category with respect to the location of the cryocooler.

Unlike with a single heat pipe tubing attached to a cryocooler recondenser, multi-tubing and non-linear type thermosiphons [12] can also be combined to connect to a liquid reservoir where the fill level of the latter is maintained by a set of cryocoolers. This will be clearer in the examples.

We can generalize as follows.

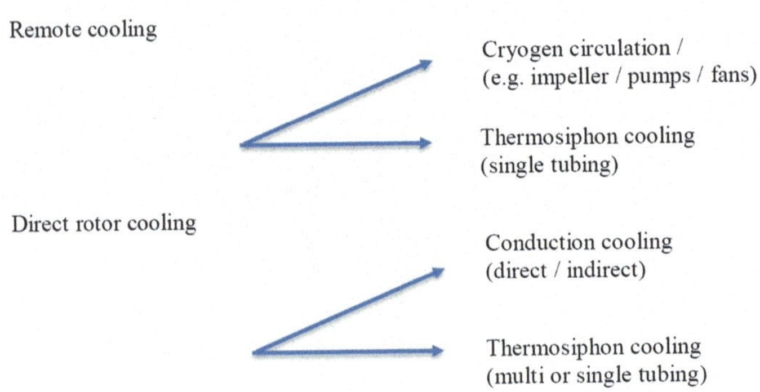

6.7 Use of Cryocoolers with SC Generator Technology (MTS/HTS)

This section gives an overview of exemplary cooling methods for promising superconducting generator technologies and designs using commercially available cryocoolers. Recently, good progress has been made in the development of high-power cryocoolers of type GM, PTR, Stirling, and other, mainly driven by HTS applications, and in particular for SC cables [19].

Figure 6.11 gives a market overview of current cooler designs, whereas Table 6.2 lists the most basic parameters for choosing the cooler type depending on the operating requirements for the respective generator design. Recently, 600 W single-stage coolers at 80 K have become available. When several of them can share one common cold box a sufficient, cold helium mass flow feeding the generator unit can be achieved. One also should note that there is a jump in cooling power from 600 W to the higher kW range for even more demanding applications, as, for example, for the SC cable industry [19].

SC generators, however, compete with existing conventional systems in the field that had the opportunity to reduce design cost over many decades. An opportunity, SC generator technology was denied of [20]. To be successful as a product and to be able to survive in a competitive market with favorable LCOE, cryogenic packaging has to be minimized. What this means is that cold box, cryocooler, and components need to be cost efficient in all operating conditions that require simple and efficient means to maintain the cryocooler as well as the complete cooling infrastructure.

Designing for modular components that can be maintained and refurbished may be the way to go, rather than the need for shutting down the complete refrigeration circuit for service.

In the following section, some typical design examples are given. As can be learned from the MRI industry, there will be a natural engineering design consolidation process due to cost pressure and the commercial availability of the individual components.

6.7.1 Overview of Available, Commercial Coolers for HTS Machinery

We now look at the landscape of current available coolers suitable for HTS applications, e.g., motors, generators, etc.

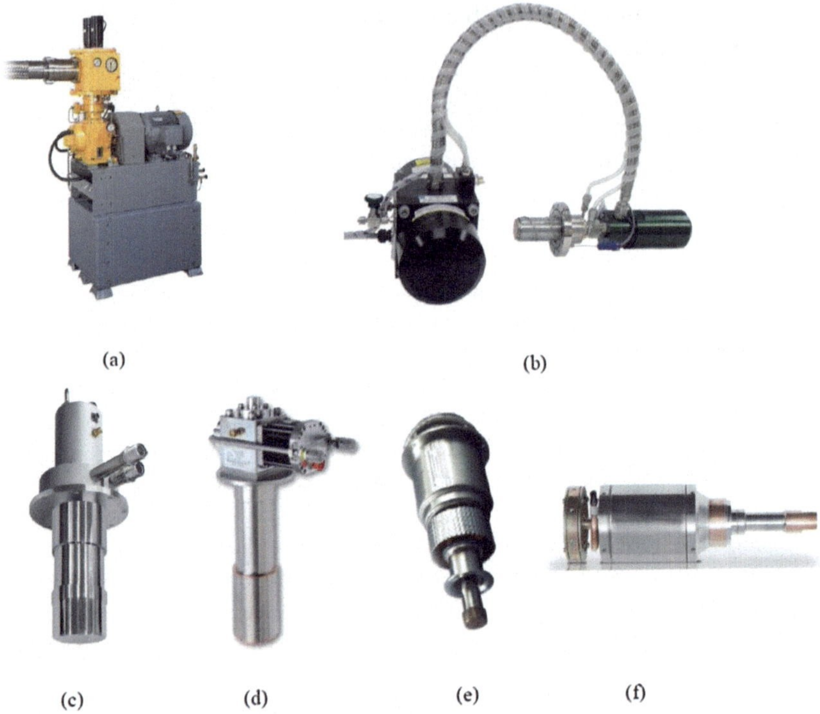

Fig. 6.11 Typical cryocooler designs suitable for HTS SC machinery

One of the systems with the longest history of operation (since 1954) is the Stirling cooler [21] by Stirling Cryogenics [22] (Fig. 6.11a). Another approach is pursued by Qdrive using a membrane driven, piston-less Stirling cooler design for the 1 kW range (Fig. 6.11b). The GM type single-stage workhorse from Cryomech (Fig. 6.11c) is now capable of supplying up to 600 W of cooling power at 80 K. In Fig. 6.11d an SHI cooler with 200 W at 80 K is shown. Recently SHI reported success in the development of a 600 W at 80 K cooler. The powerful STI Stirling cooler (Fig. 6.11e) provides up to 100 W at 80 K. Sunpower (Fig. 6.11f) produces a variety of small cryocoolers with exceptionally good MTBF and compactness. The compactness of the design is an attractive alternative for tight packaging in a cold box and modularity.

Apart from its high-cooling power, the cooler is also of interest because of its high MTBF. This is also applicable whenever cooler service is not possible during the lifetime of a superconducting unit.

The following section gives an overview of typical cooling powers.

Table 6.2 Typical cryocoolers for HTS superconducting machinery (at 60 Hz) [19]

Supplier	Type	Single stage T-range	T_{bottom} (at 60 Hz)	MTBF (h)
Cryomech	GM type	600 W at 80 K	25 K	10,000 (MTBM)
Cryomech	PTR	130 W at 80 K	9 K	25,000
Sumitomo	GM	200 W at 80 K	20 K	15,000 (MTBM)
Leybold/Oerl.	GM	140 W at 80 K	18 K	n.a.
Chart/Qdrive	Stirling	Up to 1 kW at 80 K	40 K limit	129,760
Creare[a]	Stirling	>1 kW at 80 K	(tbd.)	180,000
STI[b]	Stirling	Up to 100 W at 80 K	45 K	>1,000,000
Sunpower	Stirling PTR	Up to 16 W at 80 K	40 K limit	>1,000,000
Stirling SV	Stirling	Up to 1.3 kW at 80 K	38 K	n.a.
Linde	Rev Brayton	0.9 kW at 20 K[c]	n.a.	n.a.
Air Liquide	Turbo Brayton	50 kW and higher	35 K	n.a.

[a]Under development
[b]Design study
[c]With LN2 precool

6.7.2 *Cryocooler Cooling Powers, No-Load Temperatures, MTBF: Overview*

All coolers in the higher cooling power range of >100 W are low volume products and are therefore not shipped in large numbers. Nevertheless, the cost of the cooler is nearly negligible compared to other cost factors in a superconducting generator, e.g., in particular, the torque tube.

6.7.3 *Cryocooler Integration: Schematics and Options*

In the following, the most basic designs for the cryocooler interfacing with a cold mass are discussed. Especially for lab tests and test beds, cryocoolers already used in the MRI industry are suitable for maintaining the cooling of single race track coils at 4 K or with traditional shield coolers for HTS based coils and higher temperatures.

Figure 6.12 shows the simplest and most straightforward way of achieving a good thermal contact across the interfaces. From the supplier, the cryocooler already arrives with bolt holes on the cooling stages for bolting onto a thermal shield and/or a cold mass. Since the thermal shrinkage at the stages needs to be considered, bolts with Belleville washers are advisable in case of frequent warm up/cooldown cycles.

Directly attaching the cold stage to the cold mass interface is also beneficial for initial cooldown [23].

In the next section details of possible conduction cooling options for race track coils are given [24].

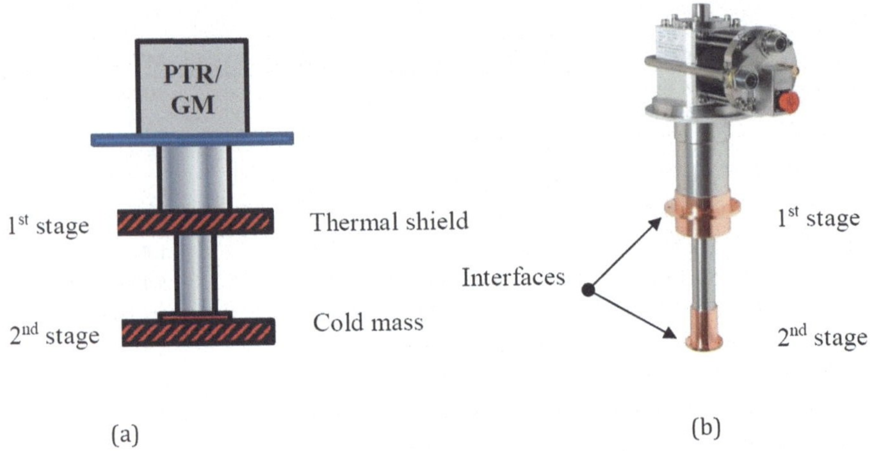

Fig. 6.12 (**a**) Cryocooler bolted onto the thermal shield and thermal mass, (**b**) SHI 4 K cryocooler RDK 408 D2 with bolt holes at first and second stages (courtesy of SHI cryogenics)

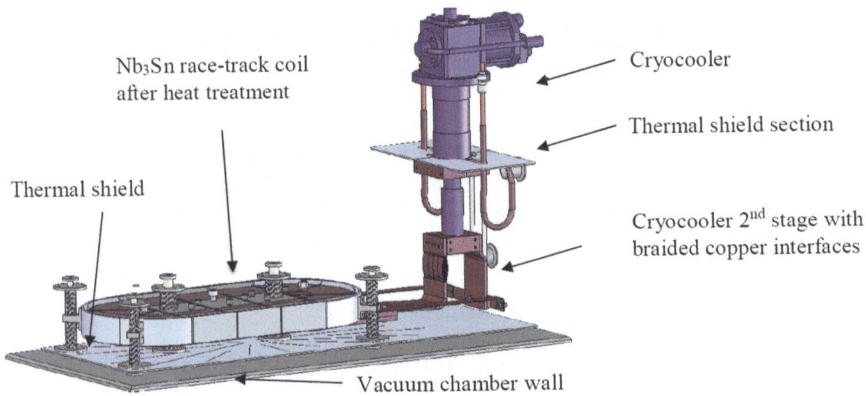

Fig. 6.13 Bolt-on design (vacuum chamber open for clarity), directly to the cold plate and race track cold mass with lead in/out

6.7.4 Conduction Cooling of Racetrack Coils

Figure 6.13 [24] shows a Nb_3Sn race track test coil cooling [25, 26] supported on Heim columns [27, 28] and with attached current leads.

For clarity, the vacuum enclosure is lifted off and only sections of the thermal shield are shown. The thermal shield is linked to the first stage of the cryocooler. The second stage is in good thermal contact with a copper cuff link and flexible braids lead into the race track coil.

Top and bottom of coil are fitted with copper plates to ensure the most distant coil part away from the heat sink is still cooled properly. The race track coil forces are constraint with welded stainless steel blocks at the circumference of the coil.

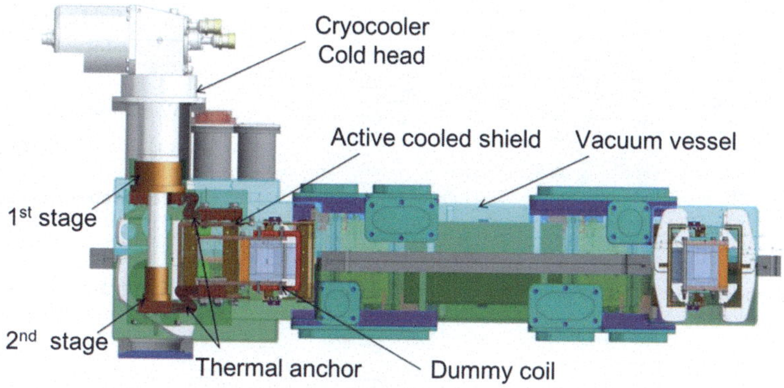

Fig. 6.14 MgB_2 race track coil for wind power applications (test bed) [31]

Recently, a similar approach has been proposed by Sun [29], [30] and the team at KIT/ITEP for testing racetrack coils for MTS wind power applications (MgB_2) as shown in Fig. 6.14.

All direct bolt-on cryocoolers may be designed sleeved or non-sleeved, depending on maintenance requirements. The same can be said for coolers that are used to recondense cryogens, e.g., helium gas.

There are downsides to both designs. A non-sleeved design requires a good 1st stage pressed contact onto the cooler interface and a workaround for any convective parasitic heat loads between 1st and 2nd stage that may arise in the absence of stratification. One also has to take precautions against air ingress that later could block internal tubing to an immersed magnet making any recondensation impossible. For the sleeved system, a good, repeatable thermal contact at both cryocooler stages is required.

Preferred contact means across the interface are either indium or the less costly Apiezon N® grease.

As with all designs, cryocooler outages and recommended maintenance schedules have to be taken into account and planned beforehand for different design options.

6.7.5 Conduction Cooling of Rotor Assembly

The following example shows the conduction cooling of a complete SC rotor (artist's view) [31] of the European Suprapower project, in which, 30 cryocoolers within the common rotor cryostat cool several single race track coils. This requires a rotary feed through for helium for the centrally placed, stationary, powerful compressor.

Note the hidden person behind (refer to arrow) that gives a size comparison, and the platform for housing the compressor, shown in Fig. 6.15.

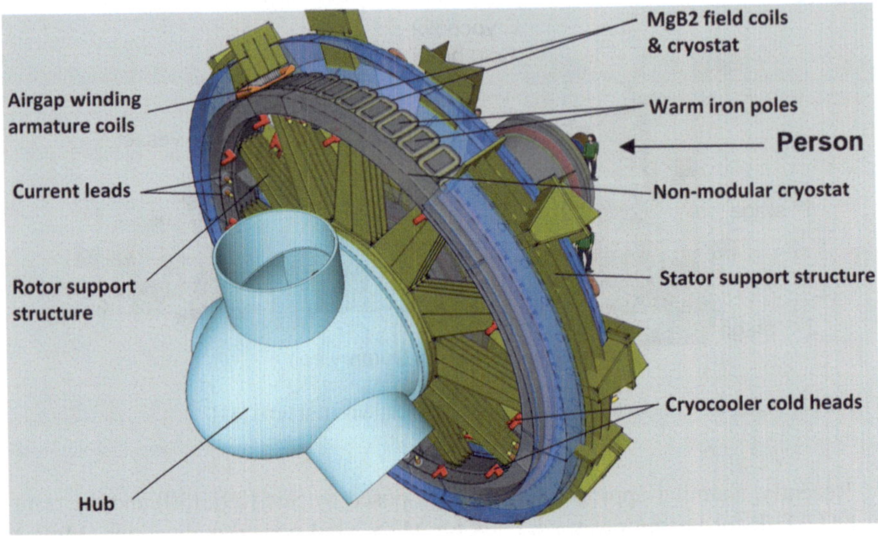

Fig. 6.15 Race-track coil for wind power applications (test bed) [31, 32]

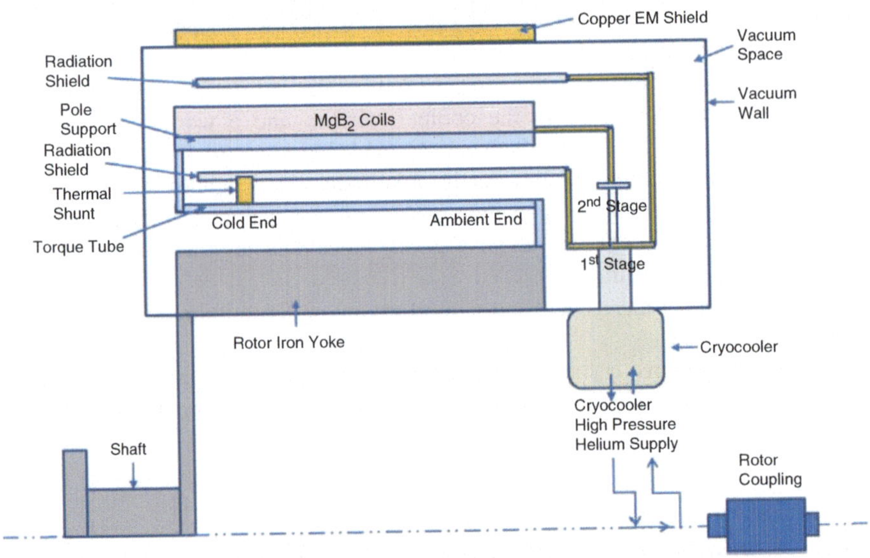

Fig. 6.16 Conduction-cooled MgB_2 coils after Kalsi [34]

Interestingly multiple coolers will be run on compressors, see also in particular the development steps given by Jakob [33].

Figure 6.16 [34] shows another direct cooler coupling approach on an MgB_2 field winding with a rotating cooler by Kalsi.

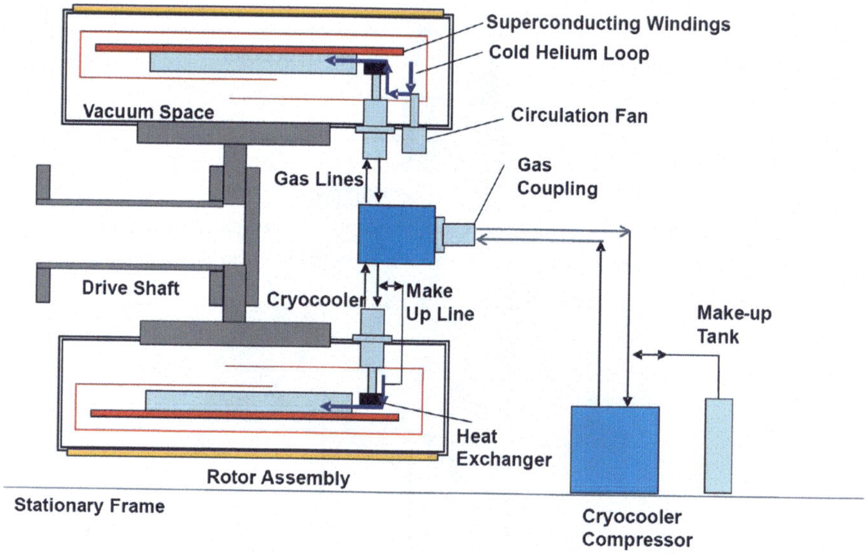

Fig. 6.17 AMSC approach (combined conduction/fan assisted cooling) [35]

Gaseous helium supplied from the compressor (not shown) flows through the rotary coupling, is fed into the cryocooler, and returned through the rotary coupling back to the compressor. Heat is being efficiently extracted from the MgB_2 race track coils and the torque tube by means of pure conduction.

Another interesting approach is shown in Fig. 6.17 as proposed by AMSC for low rotational wind speed generators of the 10 MW class at, e.g., 10–11 rpm. Here the cooler is in direct contact with a rotating HTS field winding. For longer rotor lengths a circulation fan capable of supplying a gas flow to the distant far end (x) of the SC windings and for uniformly extracting heat therefrom is required.

This also requires a thorough understanding of how the cold head performs when subject to gravitational forces over long operating hours.

As with all cryocooler integrated designs, the cooler serviceability has to be kept in mind. The temperature difference between cryocooler and superconducting winding has to be well maintained in case of a cryocooler replacement. Moreover, the cryocooler orientation dependency effect on the cooling capacity needs to be factored in into the design.

The thermosiphon cooling approach, however, allows to move the cooler away from the SC winding as discussed in the following section.

6.7.6 Thermosiphon Cooling

In 2009 GE received an AFRL contract order [7] to design a 5 MW homopolar machine for airborne applications. Figure 6.18 shows the cooling of a stationary

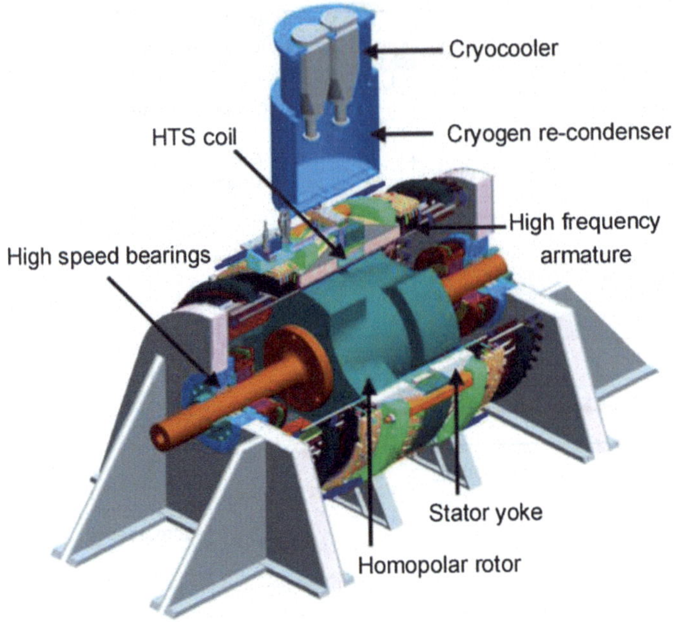

Fig. 6.18 Single-tube thermosiphon Neon cooled system with top mounted reservoir [36]

excitation coil of BSCCO-2223 type conductor with a peak field of 1 T. The rotational speed was 10,000 rpm. On top of the excitation coil assembly, a liquid neon filled cryostat was fitted that supplied liquid neon to the thermosiphon tubing. Neon boil off was liquefied with four small Stirling type cryocoolers.

A vertical, single cooling loop turn is placed around the excitation coil and filled with liquid neon. The cooling tubes lead into the bottom plate of the reservoir within the top mounted cryostat. Four cryocoolers recondense neon gas to maintain the reservoir liquid level at operating temperature of <25 K.

Many stator assemblies for different power ratings already exist in the field for conventional rotors. One would, therefore be tempted to test a superconducting generator with an already existing and stator assembly. Since the high field of the superconducting field winding rarely matches the low field of conventional stator designs, a better approach would be to design a new stator based on an optimized superconducting rotor. In any case, the strategy is to demonstrate reliability and efficiency of performance and superiority with respect to resistive systems.

Success is also reported by Siemens in 2001 using HTS conductors (10 km of BSCCO tape) with an industrial motor (97% efficiency) designed for the 400 kW range (actual performance was 450 kW) [37].

Figure 6.19a shows the non-linear thermosiphon feeder tube (shown in blue), with a small inclination angle towards the inner tube wall of the rotor. The non-rotating feeder tube (non-linear type thermosiphon) feeds liquid neon at a

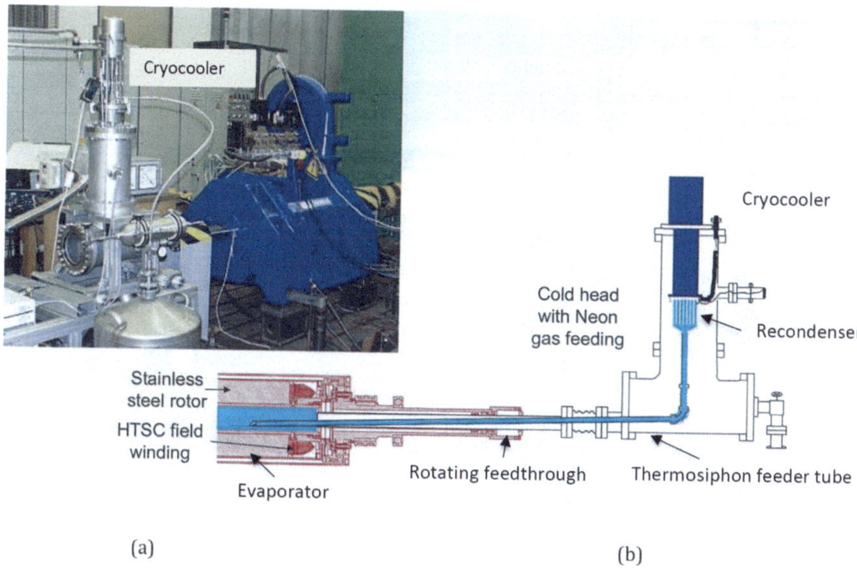

Fig. 6.19 (**a**) Neon feeder tube with recondenser, (**b**) thermosiphon feeder tube [38]

temperature of 25 K to the field winding. The neon gas boiling off and returning from the rotor is liquefied at the recondenser attached to the single-stage cryocooler [38].

As with mostly all types of thermosiphons, cooling is generally very efficient and self-regulating for variable heat loads. As a downside, the flow path from the winding to the recondenser that is attached to the cryocooler, or vice versa, should not be obstructed. A typical power requirement was 30 W at 25 K for this test system at a rotor speed of 1500 rpm.

Figure 6.20 [38] shows how the previous design concept was taken further for shipborne power generation built for high-speed applications. The stator was designed for an air core with copper Litz wire, whereas the rotor had an HTS winding (Bi-2223 pancake coils with the wire supplied by EAS), brushless excitation and a cold magnetic iron core for max efficiency. The nominal power was targeted for 4 MVA, with a voltage of 6600 V and a rotational speed of 36,000 turns/min. The reported efficiency was 98.7% [39].

In the test bed as shown in the figure above, a neon cryostat is attached to a transfer line fitted to a rotor housing. Coolant transfer to the windings is achieved again by arranging the thermosiphon tubing centrally to the inner rotor. Coolant flow is maintained by extending from a geodetically vertical height driven liquid level within the neon tank to a kinked near horizontal feeder tube that extends deep into the rotor center.

It is interesting to see how different applications can benefit from the same synergistic cooling technology approach. Figure 6.21 shows the cooling of a 10 kW

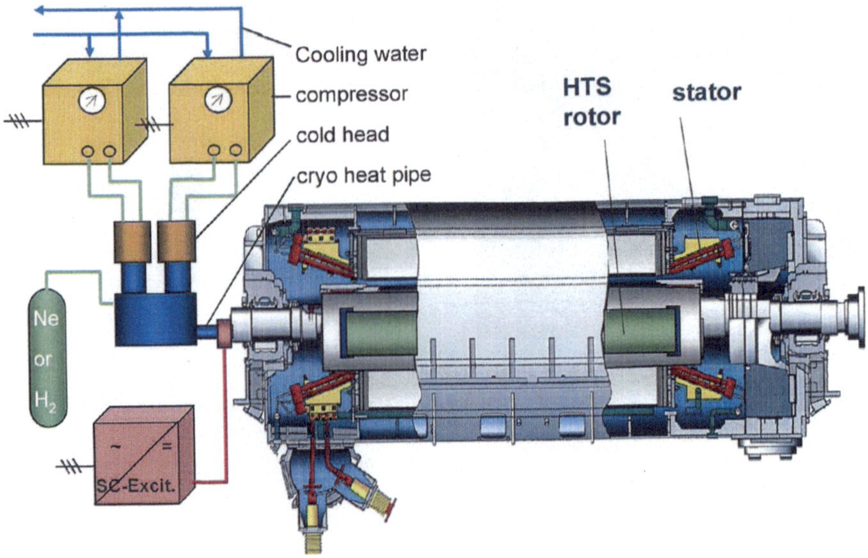

Fig. 6.20 Top mounted Neon cooled system (optional with H_2) with rotating thermosiphon, Siemens neon siphon cooling [38]

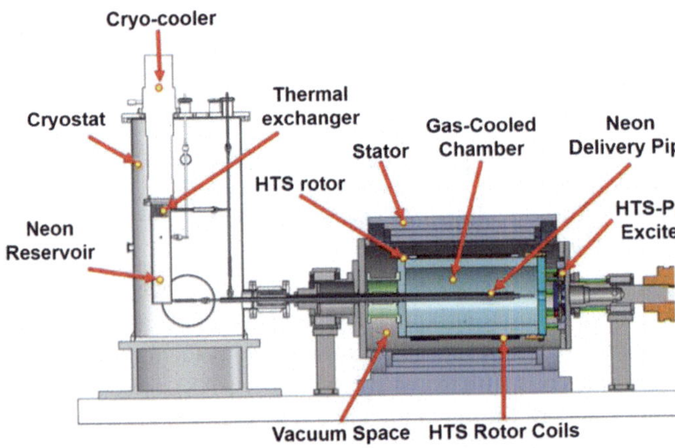

Fig. 6.21 Neon delivery tube for HTS motor [40]

HTS generator with integrated HTS-PM exciter currently under construction at Changwon National University [40]. The REBCO tape is supplied by Sunam Co. Ltd. The rotor coils are cooled with liquid neon in a closed cycle.

The horizontal neon feeder tube is non-rotational, feeding neon gas deeply into the rotor housing. 2 GM cryocoolers (1 for redundancy) were installed for supplying 40 W at 25 K.

Fig. 6.22 5000 hp synchronous motor with neon cold box [35]

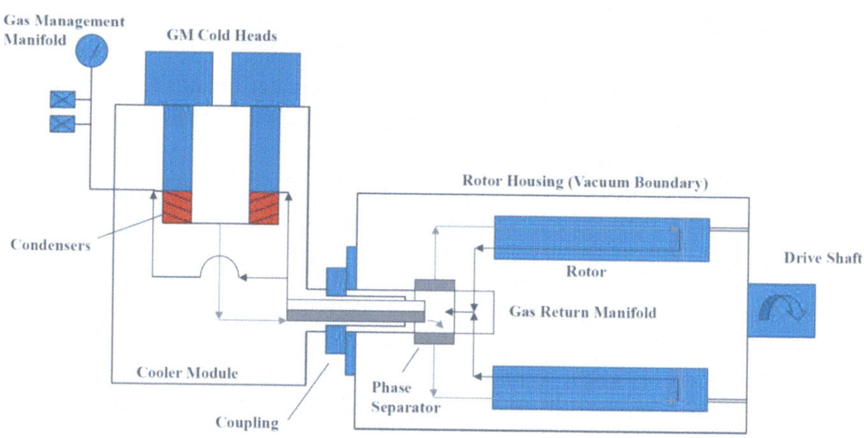

Fig. 6.23 5000 hp synchronous motor with neon cold box schematics [35]

A similar design approach was taken by AMSC for a different application, the SuperVar synchronous condenser project in 2001. Figure 6.22 shows a neon feeder tube for a 5000 hp synchronous motor with the cryocoolers fitted in a common cold box.

In this cooling version (see Fig. 6.23) two cryocoolers supply liquid neon via the thermosiphon into the rotor where the rotating phase separator drives the gas flow into rotor periphery. The flow from the field winding returns to a common manifold. The warm neon gas then returns to the cooler module (cold box) and is liquefied.

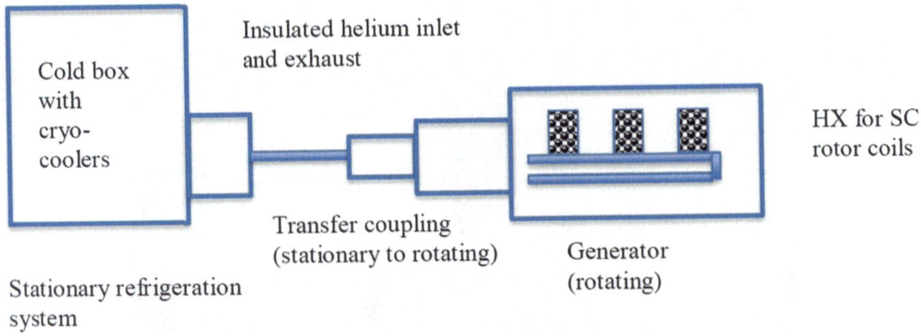

Fig. 6.24 Remote cooling building blocks [41]

6.7.7 Remote Cooling Applications

In its simplest form, the remote cooling unit consists of a separate, stationary refrigeration system. This refrigeration system is then coupled to the rotating generator via a rotating ferrofluid transfer coupling and a thermally well-insulated helium inlet and outlet tube.

Within the rotating generator, a heat exchanger is placed within the rotor cryostat (see Fig. 6.24).

Figure 6.25 shows the refrigeration cold box in greater detail [42]. Within the vacuum space a standard single-stage cryocooler is fitted with flex lines and attached to the compressor.

On the left, a high-efficiency circulator is shown. The circulator or impeller, as the name indicates, pumps gaseous helium to and forth between the cold box and the SC generator unit.

A typical cryocooler requirement would be 100 W at 40 K with a flow rate of 2–5 g/s. Apart from the cryocooler providing the cooling power, the circulator is therefore, the most important component for the cold box system.

Starting at point (1) warm helium gas pipe flow is fed into the cold box vacuum space into a cross flow heat exchanger and injected into a heat exchanger that is mounted onto a single-stage cryocooler. The flow gives up its heat to the cryocooler heat sink and flows through the connecting tubes to the generator, cooling the rotating windings. On its return path back to the cold box the helium flow passes the cross flow heat exchanger, leaves the cold box and enters the circulator. For the theory of cryogenic designs of counter flow heat exchangers, see also van Lammeren [43].

There are alternatives and different versions to this common design that allow the circulator to become part of the cold box which makes the system more compact. This can be done, e.g., if the circulator does not need to be maintained often. In other designs the cross-flow heat exchanger is omitted.

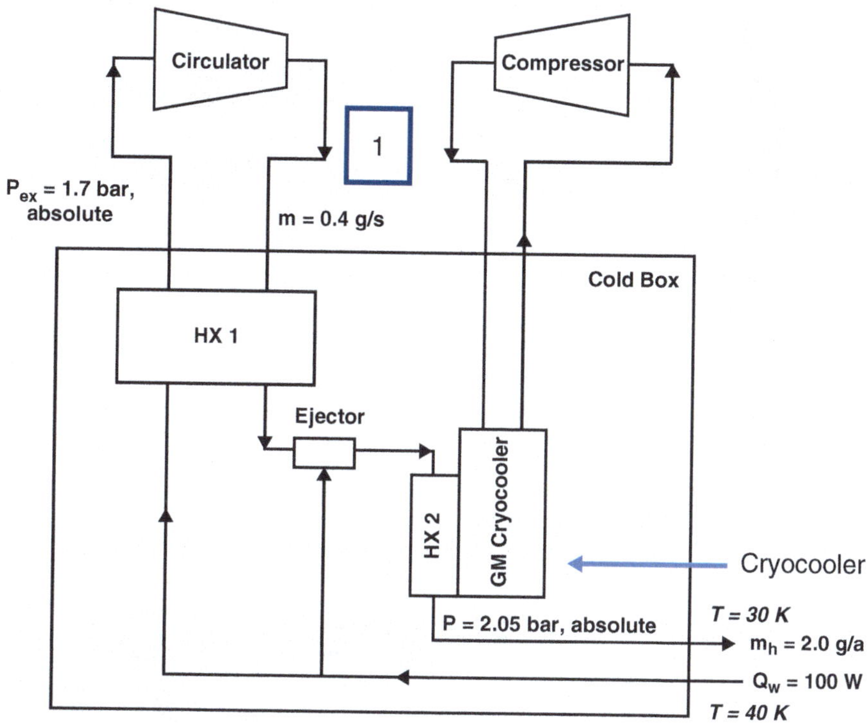

Fig. 6.25 Typical component layout for remote cooling [41, 42]

This generic type of cold boxes has also worked for many other possible superconducting applications, mainly driven by HTS technology. Examples are given by Takahashi [44] and Trollier [45].

Several novel applications have surfaced recently [19]. DH Industries USA Inc., for example, attaches a heat exchanger onto the cold stage of a cryocooler (Fig. 6.26a), mainly for steady state forced-flow HTS application cooling, e.g., superconducting cables, etc. [46]. See Fig. 6.26b.

Other circulators are available from Linde, Air Liquide, Barber-Nichols, or Mayekawa, to name but a few.

Figure 6.27 shows a cold box running with four single-stage coolers and four impellers to feed gaseous helium into the field winding of a SC generator. The transfer line is installed fixed.

Another cold box is shown in Fig. 6.28 for the Hydrogenie project of 2009 that was successfully operated and tested in 2013. This GE hydroelectric power generator was designed and factory tested to be operated for the Hirschaid Power Station for a nominal power rate of 1.2 MW [47].

The figure shows cold head and circulators in one common cold box with attached transfer lines that provided the 30 K helium fan cooling. In Table 6.3 some typical design and operating parameters are given.

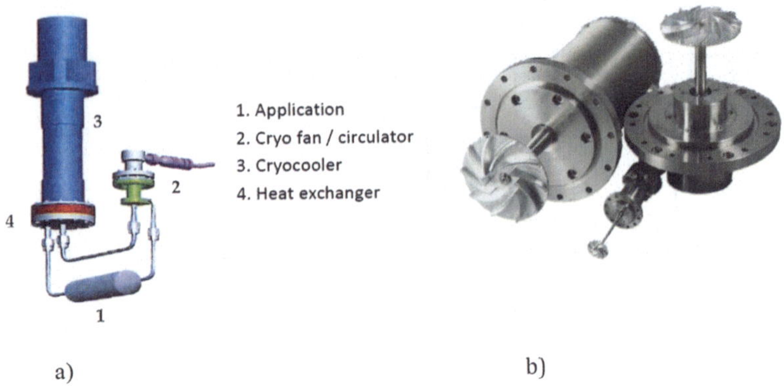

Fig. 6.26 (**a**) Cryocooler with attached heat exchanger enabling cryogen flow circulation and (**b**) cryofans (courtesy of DH Industries USA Inc.)

Fig. 6.27 Separate cold box with transfer line to generator by AMSC [35]

Figure 6.29 shows the exposed field winding (race track coils) and cooling channels that receive the impeller driven coolant from the cold box. Yet another approach has been suggested by Stirling SV and their cooler as shown in Figs. 6.11a and 6.30, as proposed for the Hydrogenie project.

This design relies on two standard Stirling cryomachines, each producing 100 W at 25 K, when feeding the flow into the rotating central tube of the superconducting rotor. The net heat load was in the range of 170 W [48].

Impellers are known workhorses with excellent reliability and MTBF.

Although upcoming pump designs are feasible and may be an interesting alternative to impellers, long-term MTBF and overall efficiency of piston driven systems versus impellers still needs to be field proven. A typical piston driven system is shown in [49] for the NEDO project (10 MW HTS wind turbine) and gaseous helium.

Fig. 6.28 Typical component layout for remote cooling, GE Energy

Table 6.3 Parameters of Hydrogenie (2009)

	Type
Supplier	
Power rating (nominal) (MW)	1.2/1.7
Zenergy conductor	Bi-2223
Rotational speed (rpm)	214
Number of poles	28
Line voltage (kV)	5.25
Design parameters	
Diameter (m)	3
Axial length (m)	1
Operating temperature (K)	30
Base temp. (with 1 cooler)	45
Coolers (2)	Cryomech
Impeller driven cooling	Cryozone

Cryogenic liquid helium circulation pumps for forced-flow cooling have been developed back in 1978 at KIT for supercritical helium [50] and also developed for ITER by KIT/ITEP (H. Neumann and G. Zahn).

In principle, those may also be suitable for cooling superconducting machinery. The ITER cooling system relies on efficient performance of helium circulation pumps. The pumps operate mainly at 4–5 K and mass flows of 2–3 kg/s.

More recently a liquid helium centrifugal pump was also developed by INFN-LNL and CERN [51].

Fig. 6.29 "Hydrogenie" assembly with racetrack coils and cooling channels, GE Energy

Fig. 6.30 Stirling cooling system layout (courtesy of Cryozone) [46]

6.8 Use of Cryocoolers with SC Generator Technology (LTS)

6.8.1 Example: Wind Energy Application—Overview

In 1974, Kullmann [52] was the first to point out and systematically investigate armature/field winding design concepts with respect to its rotation. He identified 4 possible, logical design options for a superconducting generator:

1. Inner field winding rotating/static outer armature
2. Outer field winding rotating/static inner armature
3. Inner armature rotating/static outer field winding
4. Outer armature rotating/static inner field winding

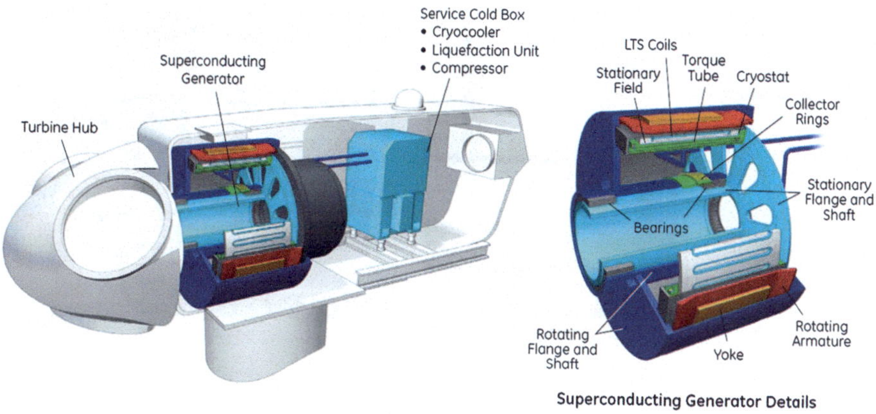

Fig. 6.31 Wind turbine nacelle with cold box

Each and every solution has its own intrinsic shortcomings. In the light of future technology efforts GE revisited designs and chose to follow approach 4. The project "Superconductivity for Large Scale Wind Turbines" was funded by DOE [53] in 2012.

Learning from the past, from rotating cryogenics, this simplified the cryogenics effort remarkably and led to some unique designs that made maximum use of the space available in the nacelle. Furthermore, it allowed a synergistic design transfer from tightly integrated cryogenic MRI systems [54] to SC machinery [55, 56].

The design approach was well received by the SC generator community [57].

Figure 6.31 shows the principal layout of a SC wind turbine nacelle and the arrangements of the individual components.

The typical field coil assembly consists of 36 individual race track coils each wound with conventional NbTi wire. The low temperature superconducting coil assembly requires a good vacuum environment with the superconducting coils operating between 4.2 to 4.9 K. Figure 6.32a and b show the intended cryogenic components and coolers for the hub end (left) and the non-hub end (right).

Since the field winding assembly is stationary in this hybrid SC generator design concept, rotating transfer couplings, forced gas flow with externally driven helium refrigerators (as mentioned in previous sections) to cope with the large heat loads are absent, and a novel cooling technology can be employed as shown in the overview given in Fig. 6.33.

The aluminum coil former supporting the race track coils is cooled by a thermosiphon cooling loop tube structure that provides uniform cooling over the large diameter and its entire surface area. Liquid helium is replenished as soon as the evaporating helium bubbles from the thermosiphon cooling loop tubes reach the recondenser of the 4 K cryocoolers.

The cryocoolers are able to maintain a constant liquid helium fill level within the tubes. For maintenance service, cryocoolers can be selectively removed from

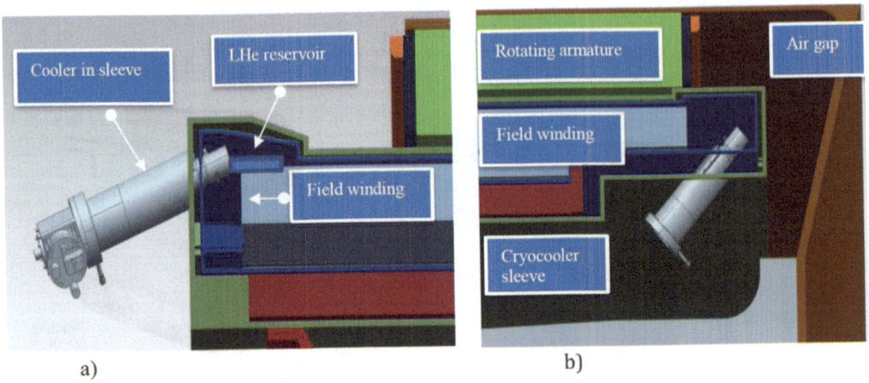

Fig. 6.32 Cooler installation at non-hub end (**a**) and hub end (**b**)

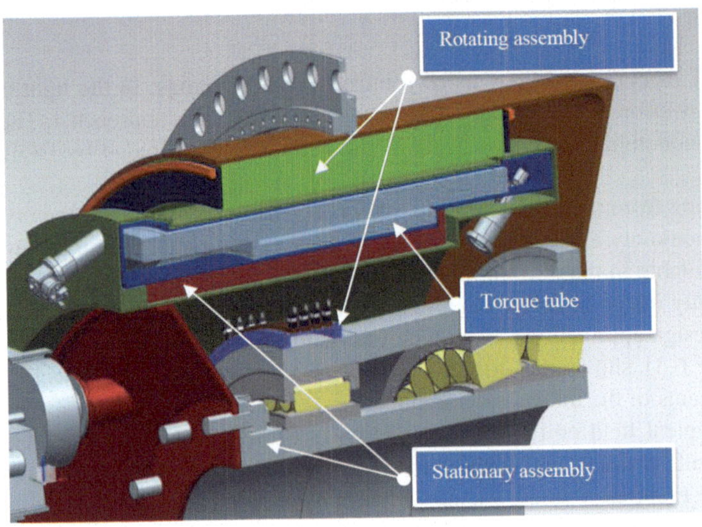

Fig. 6.33 Overview

the sleeve at field for service, if necessary, and while the armature is rotating. The time to service can be well predicted as available from standard MRI monitoring technology [58].

Table 6.4 lists early design parameters of a possible 10 MW LTS design.

The design can be scaled—with minor changes to cryogenics—for power ratings beyond 25 MW.

As of November 2019, GE has installed the world's most powerful, non-superconducting offshore wind turbine, the 12 MW Haliade-X capable of powering 16,000 European homes. This shows that superconducting technology needs to successfully demonstrate its superiority over conventional designs.

Table 6.4 Operating parameters for LTS 10 MW field coil assembly

Operating parameter		
Temperature (SC coils) (K)	4.3	4.9 max
Temperature (TS) (K)	42	45 max
Cryocoolers		
SHI RDK 415 dual-stage (4) (K) 1st stage/2nd stage	40	4.3
Cryomech AL 600 coolers (2)	Cooldown	
Stationary (incl. transients)		
Total heat loads (W) (1st/2nd stage)	125.6	4.84
Total avail. cooling power (W) (1st/2nd stage)	180	6.8
Total cryo margins available (W)	54.4	1.96
Operating time before service	>5 years	
Transients and events		
Cooldown time system (days)	$7\frac{1}{2}$	
Cooldown time TS (days)	<6	
End temperature after quench (K)	58	
Re-cool back to operating temp. (h)	21	
Ride-through (all coolers off) (h)	3	

6.9 Summary

Interestingly, many different cooling options for cooling superconducting machinery exist today, thanks to the availability of stand-alone cryocoolers and the building blocks required to assemble an efficient cold box.

The modularity of units and the variety of choices give us hope to believe that cooling of superconducting devices for the power industry will soon be a commodity with reduced and predictable risks.

We now have a choice to select coolers for different power ranges and with extremely high MTBF that should satisfy even the most demanding applications in the power industry.

However, there will not be one solution that fits all technical requirements. It is more likely a very selective and combined approach of all technologies discussed in this chapter that will make the cooling of superconducting machinery successful, relating to performance, efficiency, and cost.

On a positive end note, the cryocooler industry is listening - but we need to specify more clearly what we need and what we expect from the cryocooler performance.

For a more general overview on the development status and the technology roadmap of superconducting rotating machines including the cooling efforts using cryocoolers, see the latest topical review compiled by Haran et al. [59].

6.10 Outlook

Although coolers are known workhorses in normal climatic conditions (e.g., hospitals or similar environments [60]) no test data on the resilience of cryocoolers and compressors is available when exposing them to a rough environment and occasional cold shocks.

Temperature cycling (−40° to +60°)

1. Humidity levels (up to 95%)
2. Salinity (5%)
3. Rocking motion (±10° at 1–5 Hz)
4. Varying levels of helium purity
5. Different coldhead mounting angles during exposure
6. Rotating coldhead experiments

This is where conventional GM coolers will have to prove themselves. Alternative designs like those of Qdrive and similar will have an advantageous edge, at least for the HTS generator designs.

Last but not least, serviceability is of utmost importance. Even though performance monitoring allows prediction of coldhead and compressor failure years in advance, maintenance free, - high MTBF is key for the cryocoolers working with high performance superconducting generators, especially for offshore wind deployment.

References

1. DeSorbo W, Healy WA (1964) The intermediate state of some superconductors. Cryogenics 4:257–323
2. DeSorbo W (1964) The influence of impurities and solutes (primarily zirconium) on the high-current-carrying, high-field Nb_3Sn superconductor. Cryogenics 4:218–232
3. Buchhold TA (1965) Three-phase motor with superconductive bearings. Cryogenics 5:216–219
4. Buchhold TA (1964) Superconductive power supply and its application for electrical flux pumping. Cryogenics 4:212–217
5. Buchhold TA, Darrel B (1965) Superconductive helium pump. Cryogenics 5:109
6. DOE Cooperative Agreement no. DE-FC36-02GO11100
7. Air Force Contract FA8650-04-G-2466
8. Nusbaum M GE Power, private communication
9. Kalsi S, Weeber K (2004) Development status of rotating machines employing superconducting field windings. Proc IEEE 92(10):1688–1704
10. Andvig TA (1956) Rotating rig experiments on the effectiveness of the closed thermosiphon system for the cooling of turbine motor blades. Pametrada contract report # C108
11. Morris WD (1984) Heat transfer and fluid flow in rotating cooling channels. Research Studies Press, Wiley, Chichester
12. Lock GSH (1992) The tubular thermosiphon. Oxford Engineering Science, vol 33, Oxford University Press, Oxford
13. Mori Y, Nakayama W (1967) Forced convective heat transfer in a straight pipe rotating about a parallel axis. Int J Heat Mass Trans 10:1179

14. Hofmann A (1980) Self-regulating transfer modes of liquid helium to the rotor of a superconducting generator. Primary report no. 08.01.03P0ZA
15. Schnapper C (1978) Wärmetransport und Wärmebeürgang mit Helium in rotierenden Kanlen. Dissertation, KFK report 2654
16. Komarek P (1995) Hochstromanwendung der Supraleitung, Teubner Studienbücher, p 245
17. Hofmann A (1981) Self-regulating transfer modes of liquid helium to a rotor of a superconducting generator. Cryogenics 21:372–376
18. Stautner W, Xu M, Amm K (2014) Hydrogen cooling options for MgB_2-based superconducting systems. AIP Conf. Proc. 1573:82
19. Stautner W, Amm K, Xu M (2015) Plenary talk 1, cooling systems for HTS applications - overview and critical assessment. International workshop on cooling systems for HTS, (1st IWC-HTS), Matsue, JP
20. Stautner W (2013) Energy & power panel talk at New York state. Superconductor technology summit, Albany
21. Köhler JWL (1965) The Stirling refrigeration cycle. Sci Am 212:119–127
22. Dioguardi F, Stirling cryogenics, DH Industries BV, private communication
23. Stautner W (2017) Remote actuated cryocooler for SC generator and method of assembly the same. US 9,570,220 B2
24. Stautner W (2016) Special topics in cryostat design. In: Weisend JG (II) (ed) Cryostat design, chapter 7. Springer, New York
25. Stautner W, Sivasubramaniam K, Laskaris ET, Mine S, Rochford J, Budesheim E, Amm K (2011) A cryo-free 10 T high-field magnet system for a novel superconducting application. IEEE Trans Appl Supercond 21:2225–2228
26. Stautner W, Mine S, Sivasubramaniam K, Rochford J, Budesheim E, Amm K (2011) A cryo-free 10 T high-field magnet system for a novel superconducting application. In: Experimental results, IEEE, international magnet technology conference MT22, Marseille
27. Heim JR (1971) The Heim column, National Accelerator Laboratory. Report TM.334A, pp 1–21
28. Hartwig G (1995) Support elements with extremely negative thermal expansion. Cryogenics 35:717–718
29. Sun J, Sanz S, Neumann H (2015) Conceptual design and thermal analysis of a modular cryostat for one single coil of a 10 MW offshore superconducting wind turbine. IOP Conf Ser 101:012088
30. Sun J, Ramalingam R, Santiago S, Neumann H (2016) Preliminary test of the prototype modular cryostat for a 10 MW offshore superconducting wind turbine. In: ICEC 2016, New Delhi
31. Sun J, private communication
32. Muoz G, Merino JM et al (2012) Direct-action superconducting synchronous generator for a wind turbine. Patent application US 20120306212
33. Jakob G, Lizon JL (2010) Advanced high-cooling power 2-stage Gifford-McMahon refrigerator systems. In: Proceedings of the SPIE 7739, modern technologies in space- and ground-based telescopes and instrumentation, 77393I
34. Kalsi S (2014) Superconducting wind turbine generator employing MgB_2 windings both on rotor and stator. IEEE Trans Appl Supercond 24(1):5201907
35. Yuan J, Winn P (2015) Cooling system for HTS motors. International workshop on cooling systems for HTS, (IWC-HTS), Matsue
36. Sivasubramaniam K, Zhang T, Lokhandwalla M, Laskaris ET, Bray JW, Gerstler B, Shah MR et al (2009) Development of a high speed HTS generator for airborne applications. IEEE Trans Appl Supercond 19(3):1656–1661
37. Frank M et al (2003) Long-term operational experience with first Siemens 400 kW HTS machine in diverse configurations. IEEE Trans Appl Supercond 13(2):2120–2123
38. Nick W, Siemens CT (2005) Superconducting motors and generators. In: SCENET, Finland

39. Nick W, Grundmann J, Frauenhofer J (2012) Test results from Siemens low-speed, high-torque HTS machine and description of further steps towards commercialization of HTS machines. Phys C: Supercond 482(20):105–110
40. Bumby CW, Badcock RA, Sung HJ et al (2016) Development of a brushless HTS exciter for a 10 kW HTS synchronous generator. Supercond Sci Technol 29:024008
41. Ackermann RA, Grey DA, Funuyama S, Ito K (2003) Cryogenic refrigerator evaluation for medical and rotational machine applications. Cryocoolers 12:805–811
42. Laskaris ET, Ackermann RA (2002), Cryogenic cooling refrigeration system for rotor having a high temperature superconducting field winding and method, US6438969
43. van Lammeren JA (1941) Technik der tiefen Temperaturen. Springer, New York
44. Takahashi M, Iwai S, Miyazaki H, Tosaka T et al (2017) Design and test results of a cryogenic cooling system for a 25 T cryogen-free superconducting magnet. IEEE Trans Appl Supercond. https://doi.org/10.1109/TASC.2017.2673762
45. Trollier T, Tanchon J, Icart Y, Ravex A (2014) High capacity 30 K remote helium cooling loop. In: AIP conference proceedings, vol 1573, pp 1461–1466
46. Vermeulen H (2013) Cryogenic circulators: the solution for cooling problems? Cold Facts 29(2):49–48
47. Fair R, Lewis C, Eugene J, Ingles M (2010) Development of an HTS hydroelectric power generator for the Hirschaid Power Station. J Phys: Conf Ser 234:032008
48. Vermeulen H (2016) DH Industries USA Inc, private communication
49. Furuse M, Fuchino S (2016) Development of a cooling system for superconducting wind turbine. Cryogenics, 80:199–203. https://doi.org/10.1016/j.cryogenics.2016.05.009
50. Krafft G, Zahn G (1978) A reciprocating liquid helium pump used for forced flow of supercritical helium. Cryogenics 18:112–114
51. Pengo R, Junker S, ten Kate HHJ (2010) Liquid helium centrifugal pump characteristics from 80 g/s to 1200 g/s. Cryogenics 50:8–10
52. Kullmann D (1974) Siemens Research Labs, Superconducting generators and motors, VDI course notes Karlsruhe
53. Fair R, Stautner W (2011–2012) DE-EE0005143 report on superconductivity for large scale wind turbines. GE Global Research, NY, USA
54. Lvovsky Y, Stautner W (2013) Topical review. Novel technologies and configurations of superconducting magnets for MRI. Superconductor Sci Technol 26:093001
55. Stautner W, Fair R, Sivasubramaniam K, Amm K, Bray J, Laskaris ET, Weeber K et al (2013) Large scale superconducting wind turbine cooling. IEEE Trans Appl Supercond 23:5200804
56. Fair R, Stautner W et al (2012) ASC 2012 invited talk, Superconductivity for large-scale wind turbines, IEEE/CSC & ESAS European Superconductivity News Forum, no. 22. http://www.ewh.ieee.org/tc/csc/europe/newsforum/pdf/STP307.pdf
57. Snieckus D, Wind's Super Cool Future discussion on the GE 10 MW superconducting generator design concept, including cryogenics, published in Recharge 05/13. http://www.rechargenews.com/wind/article1293593.ece
58. W. Stautner W (2019) Cooling assembly for electrical machines and methods of assembling the same. Patent application Patent US 10,224,799 B2
59. Haran K, Kalsi S, Arndt T, Karmaker H, Badcock R, Buckley B, Haugan T, Izumi M, Loder D, Bray J, Masson P, Stautner W (2017) Topical Review, High power density superconducting rotating machines – development status and technology roadmap, Superconductor Science and Technology, vol. 30, no. 12. https://iopscience.iop.org/article/10.1088/1361-6668/aa833e
60. Stautner W, Amm K, Xu M (2016) Review on superconducting technology of MRI scanners. Plenary talk 6, In: 26th International Cryogenic Engineering Conference, International Cryogenics Material Conference, New Delhi

Chapter 7
Cryocoolers for Laboratory Research

Ravi Bains

Abstract Cryocoolers have been used in the design of dry cryostats since the introduction of the small GM-type cryocoolers over half a century ago. The advantage of using cryocoolers is to eliminate the need for liquid helium which has become more expensive, and the availability is uncertain. Liquid helium is a non-renewable natural resource and should not be wasted, some users capture the boil-off and reliquefy it, this solution conserves the helium but some helium is still lost in the process. Use of cryocoolers for research cryostats helps in the conservation of this critical natural resource. In traditional cryostats, the vibrations from the moving components were tolerated due to the nature of the bulk materials. In nanoscience experiments, the design has to isolate the cryocooler vibrations from the sample while still cooling it to the lowest temperature of the cryocooler. This chapter covers some of the principles of design used in the contemporary cryostats designed for nanoscience experiments.

7.1 Cryocoolers Used in Research Cryostats

Scientists working on characterizing material properties do so at low temperature for many reasons:

- To reduce background noise on an atomic level.
- To get sharp peaks on their optical spectra so as to be able to resolve the material composition more precisely. Broad peaks may hide some weak signals (Fig. 7.1).
- To establish transition temperature as in superconducting materials.
- To map magnetic states in novel materials. This is critical for research in topics such as spintronics and quantum computing.
- To study the crystal structures of samples.

R. Bains (✉)
Advanced Research Systems, Inc., Industrial Park Way Macungie, Pennsylvania, USA
e-mail: rbains@arscryo.com; https://www.arscryo.com

M. D. Atrey (ed.), *Cryocoolers*, International Cryogenics Monograph Series,
https://doi.org/10.1007/978-3-030-11307-0_7

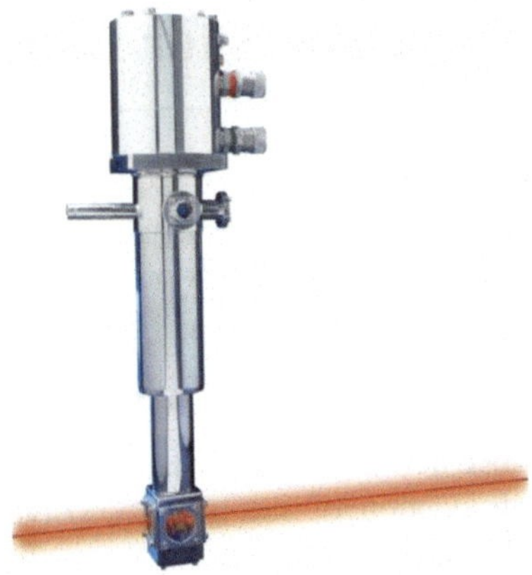

Fig. 7.1 A cryostat is a customized tool for sample environment to suit the needs of the scientist

Cryocoolers have been used as the cooling engine for "Dry Cryostats" in laboratory cryogenic systems since the first, two-stage, compact cryocoolers became popular in the 1960s. Modern day or contemporary cryostats using closed cycle cryocoolers are discussed later in this chapter as well. These cryostats have mainly been designed for material characterization and optimized for nanoscience, quantum material property testing, and 2D materials.

In this chapter we will discuss "Dry Cryostat" applications and design to overcome these limitations.

7.2 History of Dry Cryostats

Compact two-component cryocoolers have been used for laboratory research since GM cryocoolers were developed in the early 1960s. Several companies like Air Products, CTI, and RMC sold cryocoolers and cryostats for material characterization of bulk samples or thin films.

Cryocoolers were also built for cryopumping of vacuum systems. Cryopumping was an important application in the dawn of the semiconductor industry. These cryocooler temperatures bottomed out at 10 K which was the temperature required for cooling the panels of the cryopumps to pump atmospheric and process gases.

Compact, two-stage cryocoolers were made by two dominant players in the cryopump market, Air Products and CTI-Helix. Both of these companies provided cryopumps for the semiconductor markets. As an extension, both the companies also sold cryostats to scientists doing material research. Cryostats are by definition a research tool to provide a sample environment for research in a specific field of research.

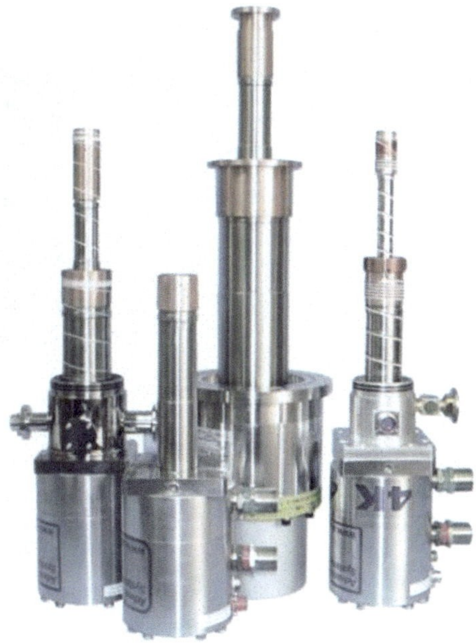

Fig. 7.2 Solvay single-stage and two-stage cryocoolers come in many sizes, temperature range, and cooling power. These are popular for use in vibration-sensitive applications

Over the years, the applications have become more complex, and the cryostats have become more sophisticated. The bulk material investigations were followed by thin film materials which were followed by surface science and investigation of 2D materials and nanomaterials.

In this chapter, we will discuss several popular applications where cryocoolers (Fig. 7.2) are used and the important design considerations to accomplish the experimental task. We will also discuss the options in design, in addition to the pros and cons of the various designs.

7.3 Elements of Cryostat Design

A typical "Dry Cryostat" consists of the following components (Fig. 7.3):

- Cryocooler, including compressor and hoses, for helium-free operation
- Experimental vacuum shroud for thermal isolation
- Radiation shield to manage radiation heat
- Sample holder for the unique experimental requirement
- Temperature control

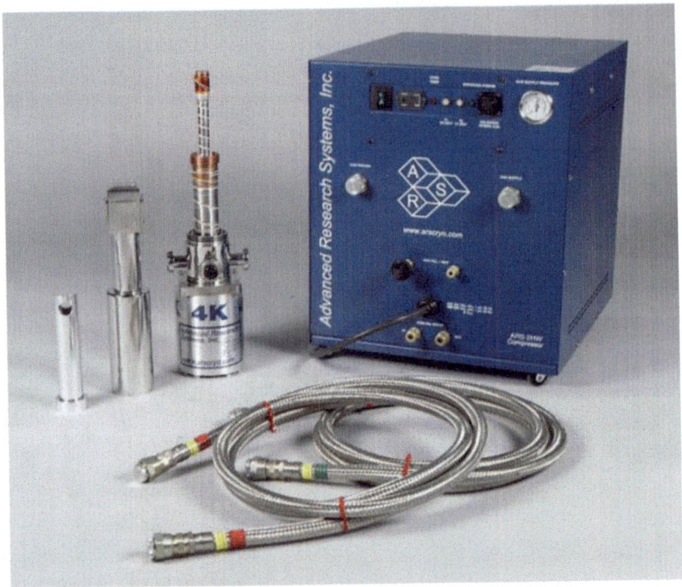

Fig. 7.3 Components of a typical cryostat. From top left, clockwise, radiation shield, experimental vacuum shroud, cryocooler, helium compressor and helium hoses (to connect the cryocooler to the compressor). ARS picture gallery

7.3.1 Cryocooler, the Engine That Drives the Cryostat

The sample temperature will depend on the cryocooler used. Cryocoolers today can achieve 10 K, 6 K, and sub 4 K temperatures depending on the regenerator bed design in the displacer. Selection of a cryocooler for a given cryostat is based on the sample temperature desired and the total heat load, experimental and parasitic, on the sample.

The sample is typically cooled by the second stage of the cryocooler, while the radiation shield and the thermal anchoring of the wiring are cooled by the first stage at 20–30 K.

The Solvay pneumatically driven cryocoolers, see Fig. 7.4, lend themselves well to the low vibrations needed for the nanoscience sample cooling. The displacer in the Solvay cryocooler is controlled by two gas buffers which control and dampen the stroke at the end of its motion in both the directions. The Solvay cryocooler has only two moving parts which are lightly loaded and result in minimal vibrations transmitted to the sample under test which is attached to the second stage.

The pneumatic drive principle uses a helium gas bubble (buffer volume) at the top and the bottom of the displacer to control the motion. By controlling the pressure of the gas at the two ends of the displacer, the motion of the displacer can be controlled very precisely. The gas bubble above the warm end of the displacer is maintained at ~200 psi. The bottom gas bubble fluctuates from ~100 to ~300 psi. The pressure

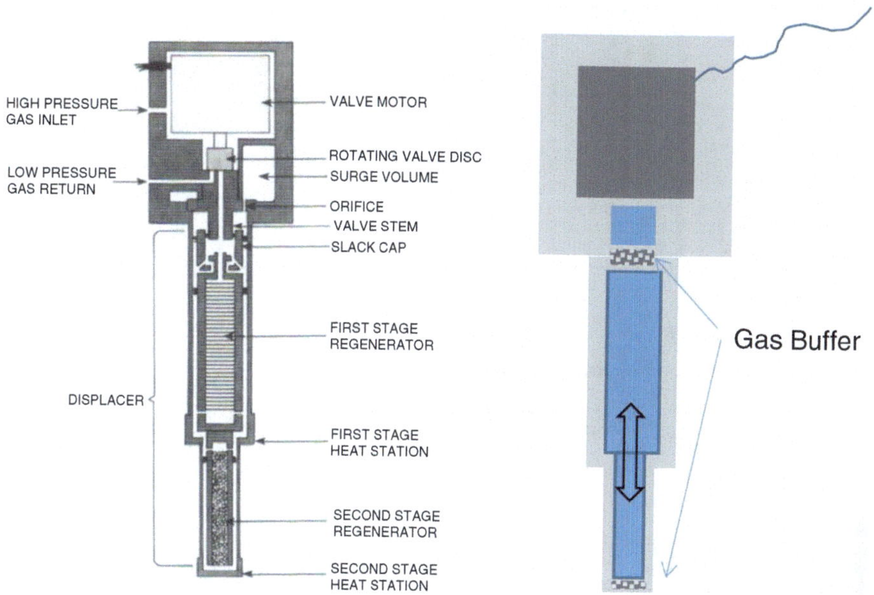

Fig. 7.4 Basic construction of a Solvay Cryocooler's unique pneumatic drive is shown on the left. The moving parts are the displacer and the rotating valve disk. Right-hand side figure depicts the helium gas bubble at the top and bottom of the displacer. There are no mechanical drive components as in the GM cooler. ARS image gallery

difference between the helium gas bubble at the top (at 200 psi) and the gas bubble at the bottom (at either 100 or 300 psi) will move the displacer in a controlled manner in one direction or the other. The regenerator material is packed in the displacer and is designed to deliver high cooling power at sub 3 K. The cryocooler can operate in any orientation as needed in cryostat applications.

7.3.2 *Vacuum Shroud Design*

All the components inside the cryostat including, the sample environment are cold and therefore they need to be insulated from the room temperature environment outside. This is done by enclosing the cold components inside a vacuum chamber (vacuum shroud). The vacuum shroud for an optical cryostat is shown in the photograph in Fig 7.5.

The reason for lower pressure or better vacuum inside the vacuum shroud are twofold:

- The lower the pressure, the lower the number of molecules floating around inside the vacuum chamber capable of contaminating the sample surface. It is desirable

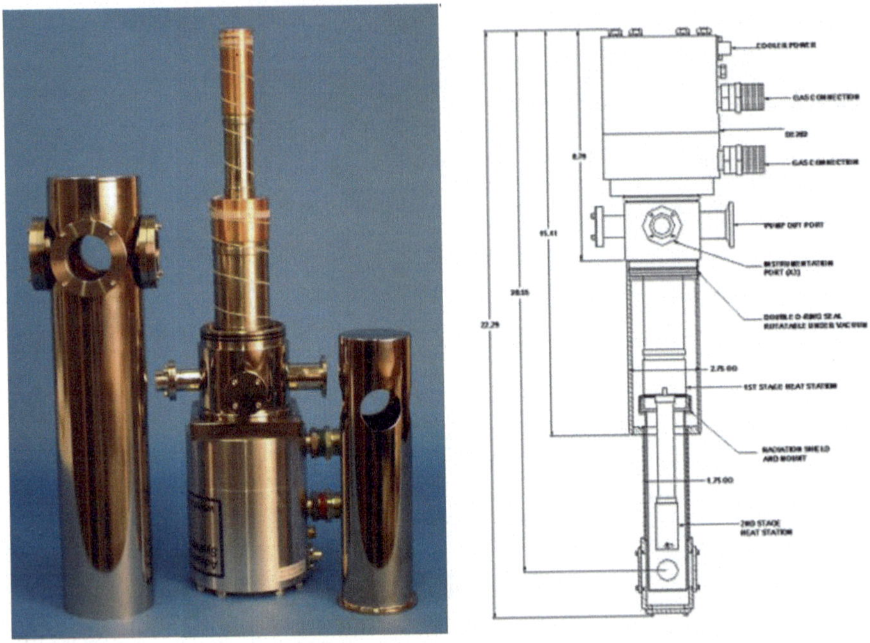

Fig. 7.5 Cryostat for optical and transport experiments. Optical vacuum shroud and radiation shield shown. ARS images

to keep the surface of the sample clean. Surface contamination can lead to errors in the experimental results.

- The insulating property of vacuum is utilized to minimize the parasitic heat load from convective losses. The higher the vacuum (lower pressure) inside the cryostat, the lower the sample temperature will be.

To get the best vacuum, it is important to follow good vacuum design practice in the chamber design.

Use UHV-compatible materials as much as possible, minimize the use of nonmetals inside the vacuum space. Nonmetals generally have a high outgassing rate. When selecting the components to go inside the vacuum shroud, it is important to check the outgassing properties of that material.

In bolted flange designs, Buna or preferably Viton O-rings are recommended. This allows the connection to be removable, and this is typical of the connection between the vacuum chamber and optical windows or electrical feedthroughs.

It is good practise to minimize the number of O-rings in the chamber for the best vacuum performance.

7.3.3 Radiation Shield Design

The radiation shield is designed to minimize the parasitic radiant heat load on the sample from room temperature surfaces. The radiation shield is cooled on the first stage of the cryocooler which typically operates at 20–30 K. The radiation shield is typically made out of OFHC copper or aluminum as both materials are good thermal conductors. The radiation shield surface can be treated to have lower thermal emissivity by nickel or gold plating.

7.3.4 Sample Holder Design

The sample holder has to serve multiple functions for the experimental scientist. It has to hold the sample in position, aligned with the optical beam. It has to get cold and have good thermal contact with the sample, so that there is very low temperature gradient between the cryocooler and the sample (Fig. 7.6).

To design the sample holder for optical experiments, the design engineer needs to ask the following questions:

Is the experiment transmission or backscattering? The answer will determine if the sample is mounted on a clear cold window (transmission) or on the copper surface (reflectance) directly.

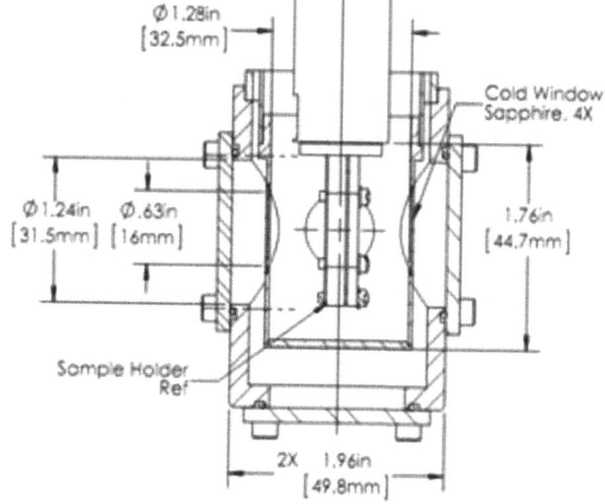

Fig. 7.6 The cross section of the shroud and radiation shield

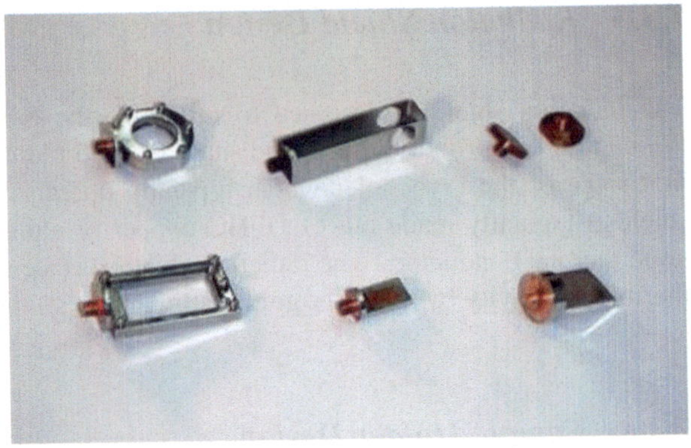

Fig. 7.7 Sample holders, clockwise from top row, left: optical sample holder for transmission. Liquid sample cell. Reflectance. Bottom row from left: large sample holder for optical transmission, powder samples, optical sample holder for reflectance or backscattering

How big is the sample? This will determine the sample space and cold window diameter.

Is the sample a solid bulk, thin film, powder, or liquid. This determines the technique to contain it in the vacuum and keep the thermal gradient to a minimum.

Some examples of sample holders are shown in the photo in Fig. 7.7.

7.3.5 Optical Window Design

The cryostat has to have optical access for the signal. This may be through an optical fiber or more often through a window installed on the vacuum shroud. The design considerations for proper window design are:

7.3.5.1 How Many Windows Are Needed for the Experiment

Only one window is needed if it is a reflection or backscattering experiment. This is also true if the inducing signal is electrical.

Two windows if there is an incident beam and a reflected beam. These may be transmission experiments or reflected experiments. The windows may be in 90° or 180° orientation.

Three windows if there is a beam splitter is involved in the experiment. The decision regarding how many windows and their relative placement is made by the optical scientist.

7.3.5.2 What Window Materials Are Acceptable

Window materials need to be matched with the wavelength of the optical signal. Quartz is the most common window material. Its transmission curve is given in the

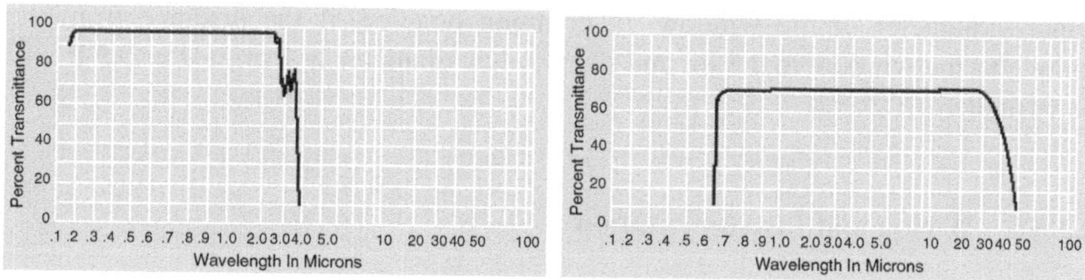

Fig. 7.8 The two graphs show the % transmission of light through the window vs. wavelength for quartz on the left and zinc selenide on the right. Quartz is the most common window material for the mid-range optical spectrum, it is the most common of windows used in optical cryostats. Zinc selenide has high transmission in the IR range. Several optical materials are available depending on the wavelength of the optical beam

fig. The transmission of quartz in the UV and IR regions is less than optimal, and there are better materials available for that (Fig. 7.8). Some of the window materials are made of salt, and these can have a short shelf life in high humidity labs.

7.3.5.3 Window Thickness Calculations

The minimum safe thickness of the window can be calculated from the formula: Thickness $= 1.1 * P *$ (window diameter)^2/Mr$^{-1/2}$

The window diameter is equal to the diameter of the supported window. The pressure P is equal to 1 atm or 14.7 psi.

The modulus of rupture, Mr is in psi. Mr of some common materials are:

Quartz = 18,000
Sapphire = 65,000
ZnS = 8000

7.3.5.4 Installation of a Window for Leak Tight Vacuum

It is important that the window be removable in the event it breaks or needs replacement. The vacuum level has to be good so the window mounting arrangement is generally an o-ring style seal. The o-ring material can be Buna or Viton depending on the level of the vacuum desired. Viton will generally result in a vacuum one decade better.

7.3.5.5 Maximizing the Optical Signal to the Sample (*F*-Number (*F*#))

The optical cone from the window to the sample has to have the largest conical angle. This is referred to as the *f*#. *F*# = distance from the sample to the window/diameter of the window. In common practice, if the distance from the sample to the window is 0.5 in. and the window diameter is 1 in., *F*# = 0.5 it will be written as *f*/0.5.

7.3.5.6 Wiring and Instrumentation

Some wiring is required for temperature control. The cryocooler will cool the sample to the base temperature of the cryocooler, but typically the research requires the sample to be programmed and tested at different temperatures, the user may want the sample to be set to a given temperature for a long period of time or may wish to ramp the temperature at some defined ramp rate. To do this, a temperature sensor and a heater is installed on the second stage of the cryocooler (Fig. 7.9). The sensor and the heater are typically wired to a temperature controller. When a set point is programmed into the controller, it reads the actual temperature, and depending on the difference between the set point and the actual, it will power up (or down) the heater to allow the sample to warm up or cool down to the set point. It is a good practice to have a second sensor to read the sample temperature independently and more precisely. Most commercial controllers have one heater and two sensor input for temperature control.

Experimental wiring is defined by the scientist and is specific to the experiment. This requested instrumentation can be in the form of:

- Copper wires for electrical signals
- Coaxial wiring for high-frequency or low-level signals
- High-current or high-voltage signals

Fig. 7.9 A thermofoil heater and temperature sensor installed. Two free length coaxial wires can also be seen for signal input and output. ARS photo library

7.3.5.7 Copper Wires for Electrical Signals

Copper wires are commonly used for small signals. Electrical or transport experiments require wires of small diameter generally between 30 and 40 AWG. Copper wire should be insulated by Teflon, Kapton, or other vacuum compatible insulation.

7.3.5.8 Coaxial Wiring for High-Frequency or Low-Level Signals

Coaxial wiring is used for low noise signals where the signal needs to be shielded from the RF noise. The conducting wire is shielded. The shielding is grounded. This allows very low experimental signals to be transmitted from the sample to the outside of the cryostat.

7.3.5.9 High-Current or High-Voltage Signals

Special wiring is available for high-current and high-voltage experiments. The high current required the copper conductor to be of a large diameter which will conduct a larger amount of heat from the room temperature feedthrough to the sample. High-voltage wiring generally has a high thickness of the insulation around the conductor. Thermal anchoring of both these types of wires to the first stage is a very important part of the design.

7.4 Design Considerations of Cryostats

Good design of a contemporary cryostat for nanoscience requires that the following features are incorporated into the design:

7.4.1 Thermal Budget

The thermal gradient from the cryocooler to the sample must be kept to a minimum. This low "thermal budget" mentality requires meticulous design from a thermodynamics point of view. In a typical nanoscience cryostat, the sample under test is small, it requires very little cooling power to cool it, and the lower the sample temperature the greater the ability to study the properties at a wider temperature range. Most of the cooling power from the cryocooler is used to overcome the parasitic losses along the path from the cryocooler to the sample. A low thermal budget philosophy requires a design that minimizes these losses from parasitic heat loads.

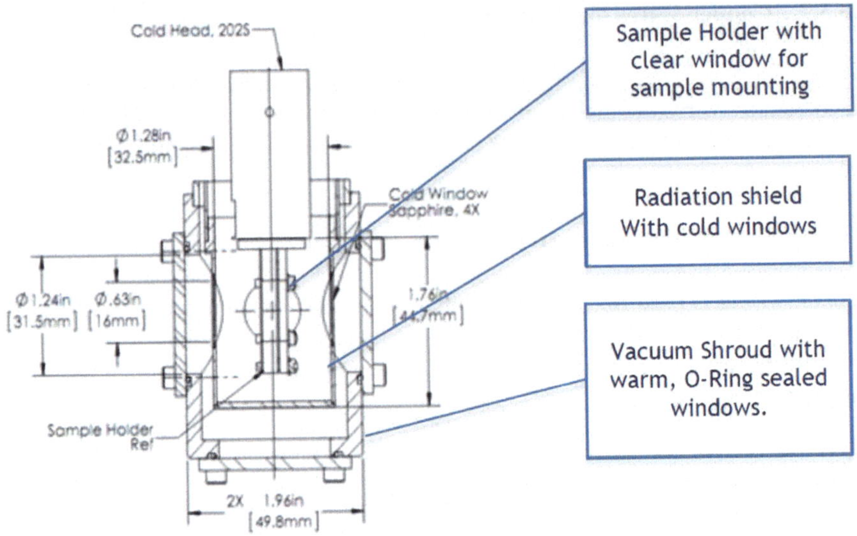

Fig. 7.10 The sample holder, the radiation shield with a cold window, and the vacuum shroud with bolt on windows

Research in nanomaterials is generally done in sub 4 K temperature region; since the modern two-stage cryocoolers can easily operate in the sub 3 K range with good design, the sample can be within a few tenths of a Kelvin of the base temperature of the cryocooler.

All components have to be very high thermal conductivity at 4 K. All joints have to have extremely low contact resistance. It becomes vital to select the correct materials for high thermal conductivity and make design provisions for minimal thermal radiation and convection loads to the sample.

For minimal sample temperature, it is also necessary to minimize the heat load from the instrumentation wiring, both parasitic and experimental. This is done by using low thermal conductivity material such as phosphor bronze wherever possible and wire of small diameter. All the wiring should be thermally anchored to the first stage of the cryocooler at 30–40 K.

The application requires optical access; a minimal opening should be allowed, and use of cold windows on the radiation shield is a good practice, see Fig. 7.10.

7.4.2 *Cryostat Construction: Material Selection*

All materials used in cryostat design have to be carefully selected for performance. The cooling or conductive path has to be a material of high thermal conductivity at the cryostat-rated temperature. Typical material for good thermal conductivity is OFHC copper. This can be further improved by selecting a high-purity copper and then heat treating it. In the design you may have multiple components joined together to form the thermal path to the sample. Fig. 7.10 shows the cryocooler and the sample holder with the sample mount (window for sample mounting).

In this setup, the following design considerations are taken:

The sample holder should be manufactured from high-purity copper, preferably out of one piece. If it is manufactured from multiple pieces, it must be vacuum brazed for minimum thermal gradient across it. A polished surface will reflect the radiant heat.

Contact resistance between the sample holder and the cryocooler should be minimized by high contact force, and a suitable filler material such as Crycon Grease (copper-loaded vacuum grease) or an Indium gasket must be used. Another solution is to gold plate the sample holder and then use high contact force. The soft gold surfaces make better contact because there is an inherent give in the material. High contact force can be achieved by using screws to either clamp the sample holder to the cryocooler or install one large screw thread to the sample holder itself. The thread surface acts as a high contact force between the cryocooler and the sample holder.

The next step is to design the sample holder window for transmission spectroscopy experiments. The window should be made of a material with high transmission in the wavelength of interest to the experimental scientist and also have high thermal conductivity. If possible, it should be made out of sapphire or quartz both of which have a high thermal conductivity and a fairly good transmission for visible light spectrum.

To get good mechanical contact between the sapphire window and the sample holder, an indium gasket is machined into the sample holder. Indium flows when the two components are assembled tightly together and provide a good thermal path from the cryocooler to the sample.

7.4.3 Thermal Resistance Considerations

The objective of the thermal design is to minimize the thermal gradient from the cryocooler to the sample under test. The sample must have the widest temperature range possible.

7.4.3.1 Thermal Conductivity

In all materials, the thermal conductivity peaks around 20–40 K and drops off quickly below 10 K. To minimize the thermal budget lost in the design components, choose the materials with the best thermal conductivity at and below 4 K. Typically this means that the components are made from very high-purity copper and annealed for maximum thermal conductivity performance at 4 K. If these cryogenic components are to be bolted together, they should be gold plated to reduce the emissivity as well as improve the contact resistance.

7.4.3.2 Thermal Contact Resistance

This is explained in Figs. 7.11 and 7.12. Any surface when seen under a microscope has high and low spots. Any two flat surfaces will only touch at three points, and all other parts of the surface are separated by a small gap. Lapping will help but not eliminate the problem. The gap must be filled by a soft conductive material or the contact force has to be high.

7.4.4 Vacuum Considerations

The higher the vacuum inside the chamber the better the thermal insulation from convective thermal losses. The added benefit of a high vacuum is the lower the number of molecules inside the sample space the cleaner the sample surface will be which is a highly desirable specification for a cryostat.

Fig. 7.11 Contact resistance at a mechanical joint is due to the surface roughness on a nano scale. The gaps need to be filled either by a filler material such as indium or the contact force has to be very high

Fig. 7.12 Contact resistance is inversely proportional to the normal force applied. The photo shows a flanged connection with multiple screws to increase the normal force and the thermal contact

7.5 Design Considerations of Cryostats for Nanoscience

The new generation of contemporary cryostats designed to study materials for nanoscience poses many challenges for the design engineer above and beyond the typical cryostats. Due to the nature of nanoscience research, the vibrations and drift control have to be managed in the nanometer level.

All closed-cycle cryocoolers have inherent vibrations from the moving components of the cryocooler as well as the "balloon effect" from pressurization and de-pressurization of the gas in the cryocooler cylinder. In the past few years, the research in nanoscience has advanced the field of medicine, quantum physics, electronics, optics, and magnetism. More recently, the focus of research has been in nanoscience; this area of research has the potential to deliver quantum computers for artificial intelligence, quantum transmission for military communications, and other evolving applications in the computers, defense, and medical fields. Cryostats for these vibration sensitive applications are shown in Figs. 7.13 and 7.14.

7.5.1 Ultra-Low Vibration Cryostats

Sample vibrations have to be in the single-digit nanometer level in cryostats designed for nanomaterial property testing. Some of the applications where sample vibrations are critical include:

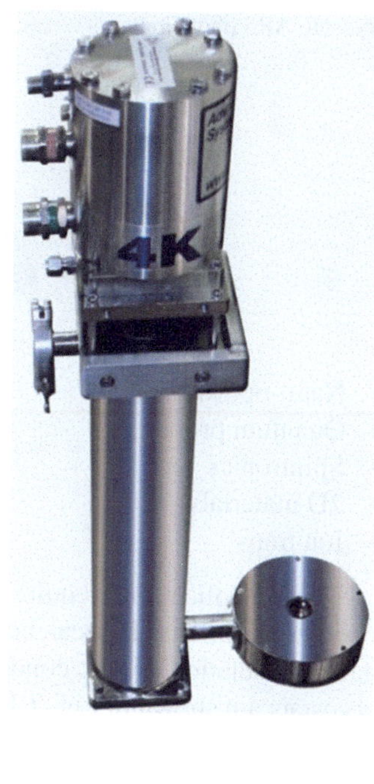

Fig. 7.13 X20 cryostat for nano-optics. The cryostat design is for a high numerical aperture for near-field experiments. ARS website image gallery

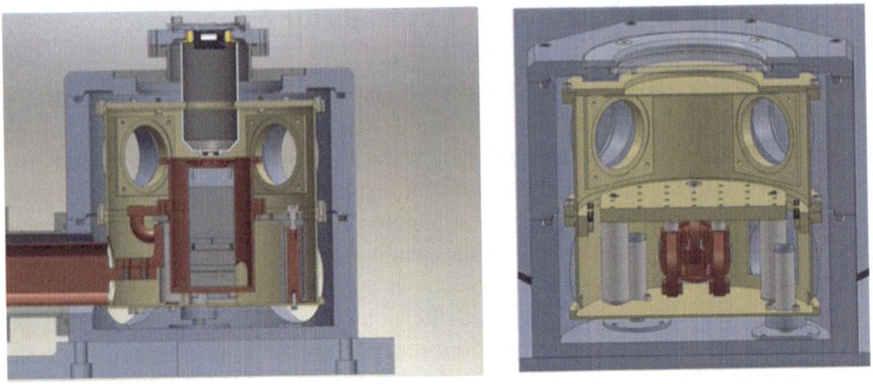

Fig. 7.14 X23 cryostat for nano-optics. The cryostat design is for a high numerical aperture for near-field experiments. In the figure on the left, the objective is seen inside the vacuum space to minimize the distance from the sample to the objective. The sample plate is supported on an ultra-low expansion support to minimize any drift during temperature ramping of the sample. ARS drawing

Fig. 7.15 The cryostat, X23 is designed for nano-optics and nano-transport experiments. This cryostat was designed for applications requiring low vibrations and drift over a wide temperature range. Figure 7.14 shows the internal construction of this cryostat. ARS photo gallery

- Nano-optics
- Quantum properties
- Spintronics
- 2D materials
- Ion traps

These applications require the sample under test to be isolated from the cryocooler vibrations. This can be done in several ways; the most common methods are gas gap isolation or soft conductive coupling such as copper braids or foil. These cryostats are shown in Fig. 7.15.

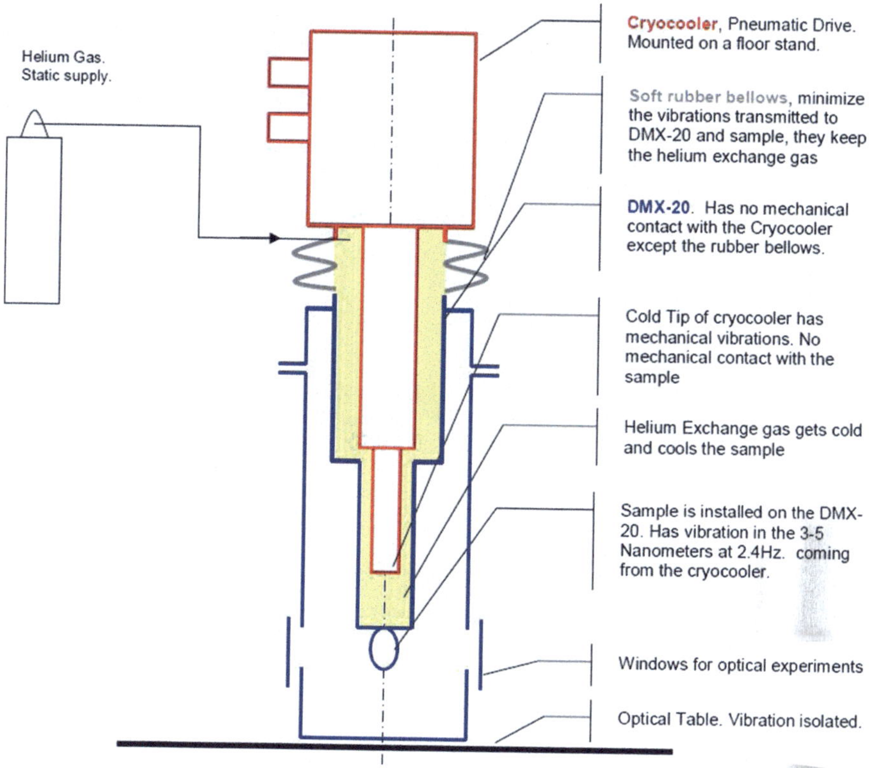

Fig. 7.16 Gas gap vibration isolation principle

7.5.1.1 Gas Gap Vibration Isolation (ULV and XLV)

In this technique, the cryocooler is positioned inside a closed cylindrical interface (ULV interface) with a slightly larger diameter than the cryocooler. The space between the cryocooler and the ULV interface is filled with helium exchange gas at atmospheric pressure. To keep the helium from leaking out of this space, the connection between the cryocooler and the interface is completed with a soft rubber bellows. The bellows prevent any vibrations from the cryocooler to be transmitted to the sample. Helium acts as the thermal path from the cryocooler to the sample (Fig. 7.16).

Helium has a stratification property. When the cryostat is in a vertical position, the coldest helium settles at the bottom which is also the coldest part of the cryocooler, and the warm helium stays static at the top; there are no convective currents. This cryostat works only when the cold tip is down. Typical vibrations at the sample will be in the 3–7 nm (peak to peak displacement) at the frequency of the cryocooler operating frequency (2–2.4 Hz). Vibrations at higher frequencies the displacement is generally lower as can be seen in Fig. 7.17.

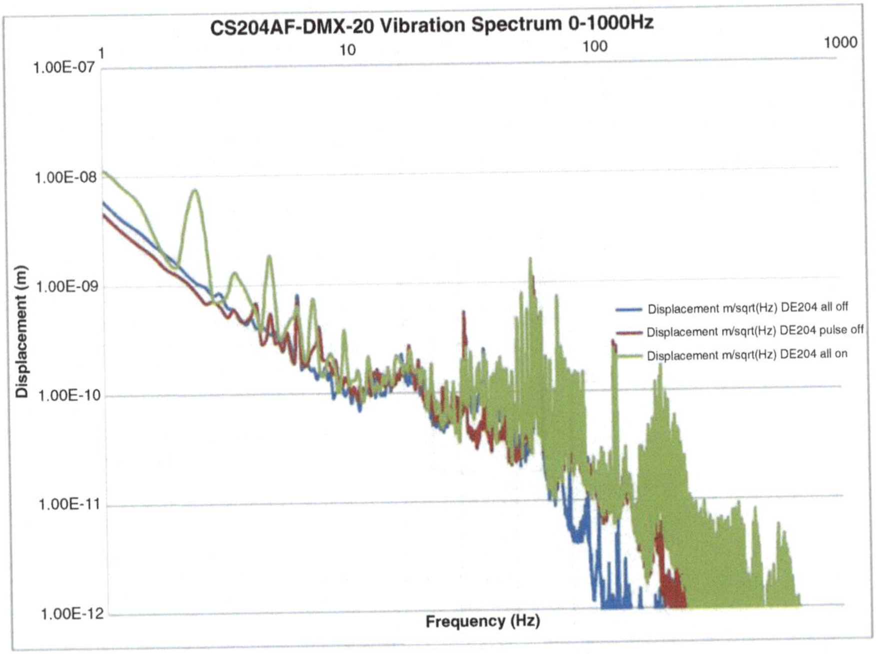

Fig. 7.17 The vibration spectra of the gas gap damped cryostat shows that the peak at 2.4 Hz is the highest between 5 and 7 nm (1.00E−8 m = 10 nm). The higher frequencies are lower than 1 nm

7.5.1.2 Soft Coupling Vibration Isolation

This technique uses a very flexible copper braid (or a thin, flexible copper foil) to isolate the vibrations at the sample, and it is less complex and more popular. The cryocooler can be operated in any orientation. The thermal conductivity between the sample and the cryocooler can be controlled by using more or less of the soft copper braids following the rules of the equation for thermal conductivity. The heat transfer will be proportionate to the area of cross section of the combined braid wire and inversely to the length of the braid. Figure 7.18 shows the flexible braids connected to the bottom of the cryocooler, and Fig. 7.19 shows the vibrations at the sample. Although the vibration will be a little bit higher, the thermal gradient between the sample and the cryocooler will be lower.

7.5.1.3 Drift Control

In nano-optics, scientists are studying the quantum properties of very small features (quantum dots) on the nanomaterials such as 2D and nano wires. The signals coming from the sample can be weak to the extent of single photon counts. This necessitates the ability to capture the emission from a small feature for extended periods of time

Fig. 7.18 Flexible copper braids shown installed at the bottom of a cryocooler for vibration isolation at the sample. The braids are covered by MLI to minimize heating from radiation. ARS photo library

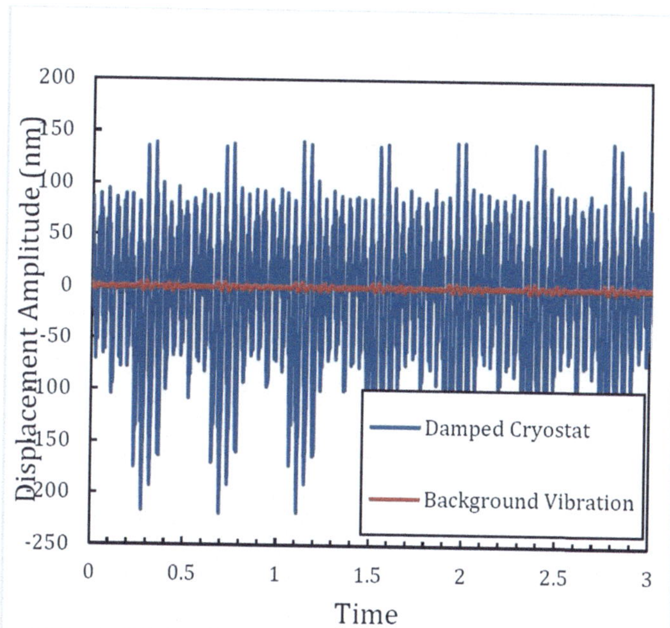

Fig. 7.19 Vibrations at the sample for the soft coupling vibration isolation method. The braids will transmit slightly higher vibrations, but the thermal gradient will be lower. The background vibrations are in the range of 3–5 nm; the rest comes from the cryocooler itself. ARS test data

sometimes over a period of several months. Any drift which may misalign the spot of interest and the detector is unacceptable. To minimize this problem. Low CTE materials can be used as structural components. Low CTE materials include Kovar and Invar. The sample mounting plate is supported by the low CTE materials, and the structure is designed for compensating thermal expansion to provide a null effect at the sample.

7.6 Popular Applications for Low Temperature Research

Typical research can involve optical, electrical, or magnetic property testing, or all three simultaneously. Optical spectroscopy can include UV, Vis, or IR. The samples can be solid, liquid, or powder.

More specialized beams may be used from synchrotrons and neutron scattering beam lines, or they may be terahertz, positron, X-ray, and neutron from sources available to the scientists in their laboratory. Knowledge of these applications is important because they all have specialized design requirements important to the design of the cryostat.

There are many considerations for the design of cryostats used for scientific research including nanoscience (Fig. 7.20).

Questions to consider during the design phase include:

- What are the thermal properties of the sample under test (SUT)?
- Is the SUT thermally conductive?
- Is the SUT a solid, liquid, or powder?
- Is it fragile and delicate? Does it get damaged easily?
- What is the temperature range for the experiment?
- Can the sample be epoxied or glued to the sample holder?
- What is the duration of the experiment? How quickly does the sample change need to take place?

Fig. 7.20 The internal structure shows the construction of a cryostat for nanoscience. The breadboard sample plate allows a lot of flexibility for the scientist to add optical components close to the sample

- How much vibrations are acceptable between the sample and the objective? How much drift is acceptable, in say, nm/h?
- Is mapping or scanning the sample surface needed?
- Do the samples require a higher vacuum to avoid contamination of the surface?
- Does the scientist need a high numerical aperture for weak signals or single-photon collection? Does this mean an objective must be placed inside the vacuum space?
- Does the user want XYZ translation of the sample? How much travel does he/she need?

7.6.1 Optical Spectroscopy

Optical spectroscopy can range from the study of bulk sample to thin films to nanomaterials. Each of these require a unique cryostat design. The current trend is towards nanomaterials, and in this chapter we will concentrate on this (Fig. 7.21).

In the high-performance cryostats designed for quantum materials or nanomaterials, the property testing typically requires the following features:

- Low sample temperature. Typically below 4 K.
- Low vibrations, 3–5 nm.
- Low drift of the sample during extended experimental runs.

High NA optics, small distance from the objective to the sample, Windows for the optical beam. Windows should be coated as required by the experiment.

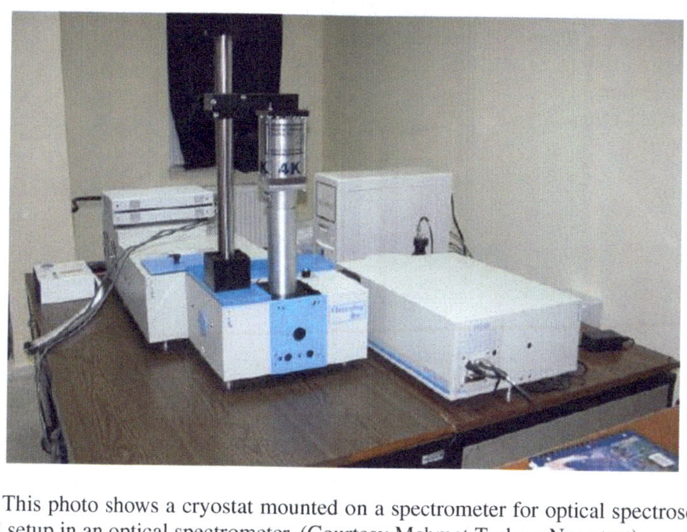

Fig. 7.21 This photo shows a cryostat mounted on a spectrometer for optical spectroscopy. This is a typical setup in an optical spectrometer. (Courtesy Mehmet Turken, Normtest)

7.6.2 *Electrical Transport*

Typical cryostat design for an electrical experiment will require a specialized sample holder for holding a dual in-line chip (DIP) or an LCC or some other holder. The sample holder must incorporate the electrical package. Keep the device cold and have the ease of use to simply plug the sample to the measurement instruments outside the vacuum environment.

The cryostat can be a sample in vacuum or a sample in exchange gas (top loader) design shown in Fig. 7.22.

The cryostat design must take the following into consideration: DC or RF signal to and from the sample.

Low-voltage signals, special care to minimize the noise from external sources, high-voltage signals, special wiring with high insulation to prevent arcing, low current, requires shielded wiring, coax or triax, high current, thermal anchoring, to minimize the heat leak from 300 K environment. Plugs and receptacles should be installed between the sample holder and the cryostat so it can be worked on outside; this makes a user-friendly operation.

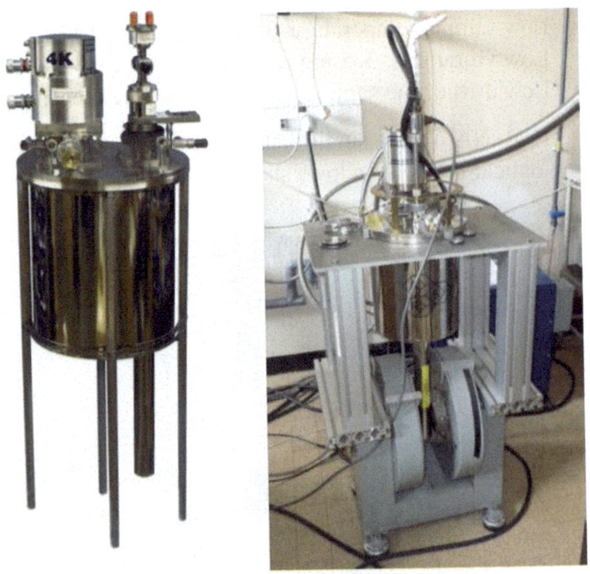

Fig. 7.22 X19 Cryostat (top loader) for electrical, magnetic, or magneto-transport properties. The sample tail can be inserted into the warm bore of a magnet, and all electrical wiring is introduced from a room temperature hermetic electrical feedthrough. Photo on the right. (Courtesy Dr. Laurent Vila, Institut Nanosciences et Cryognie, CEA-Grenoble)

7.6.3 Magnetic Property Testing

A low vibration cryostat can be designed for only magnetic property testing or it may be for magneto-optical or magneto transport research. The cryostat can be designed with the necessary features.

The top loading cryostat in Fig. 7.22 shows a sample tail where the sample can be placed. The sample is cooled by a column of He gas which is cooled by the cryocooler the DE210. Electrical wiring can be installed in the cryostat for experimental testing. Typically the sample is in a helium vapor in this type of a setup.

7.6.4 Cryostat for Electro-Optical Experiments

In this experimental setup, the sample will have electrical connections as well as optical access through a window. Leadless chip carrier (LCC) sample holders are often used with wiring thermalized at the first stage of the cryocooler.

The wiring can be copper wire and/or coaxial wires, and the feedthroughs have to be hermetic and leak-tight. The wiring is thermally anchored on the first and second stages. The important consideration is that the end of the wire connected to the sample is at the same temperature as the sample. It is important to avoid any signal distortion due to the Seebeck effect (Fig. 7.23).

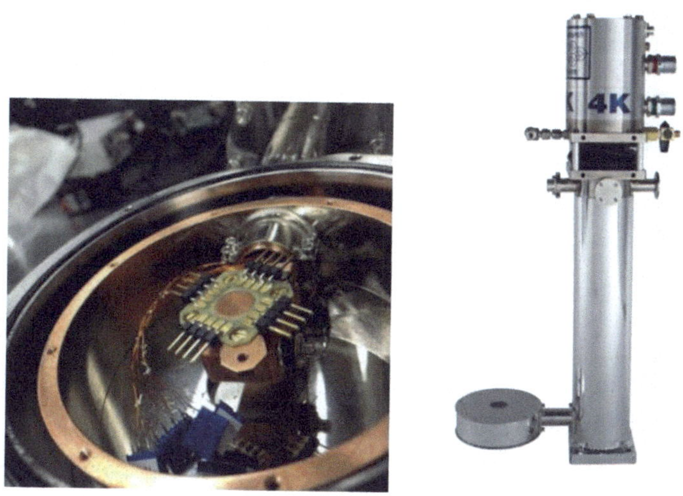

Fig. 7.23 The cryostat for electro-optical experiments. The sample can be a flat pack as shown, a DIP or a leadless chip carrier

7.6.5 *Cryostat for Magneto-Optical Experiments*

These applications include nanoscience and MOKE experiments. The cryostat is designed to insert the sample compartment of the cryostat into the poles of an electromagnet (EM) or into the bore of a superconducting magnet (SCM) (Fig. 7.24).

7.6.6 *Cryostats for UHV and XHV*

Ultra-high vacuum (UHV) is a common requirement for research where the surface quality of the sample is key to good science. Extreme high vacuum (XHV) is a relatively new requirement by scientists, making the quantum devices where the presence of any stray molecule of air or any contaminant can contaminate the surface of the device on a molecular level. These cryostats for these applications are cryocoolers with all welded joints and all metal seals such as conflat flanges. A very special requirement of this application is the need to maintain the XHV vacuum while the cryocooler is in either maintenance or repair. Breaking the vacuum to remove the cryocooler is not an option. The Solvay cryocoolers lend themselves

Fig. 7.24 The cryostat for magneto-optical experiments. The electromagnet or an SCM can be inserted over the sample tube. The sample can be mounted on a Piezo XYZ stage as shown on the right. ARS picture library

well to this XHV application because the moving parts of the cryocooler can be replaced easily without breaking vacuum.

In the construction of these cryocoolers, special attention is given to the materials of construction including the grades of stainless steel for the cylinders, the flanges, and all the copper parts used. Manufacturing procedures have to be modified to ensure that brazing and machining do not result in contamination inside the vacuum because of either the cutting fluids used or the flux.

Frequent baking of the cryocooler along with the chamber at over 500 °C for several weeks is also a requirement (Fig. 7.25).

7.6.7 Cryostats for Surface Science

Surface science is the study of the properties of surface or the interaction between two surfaces of a material. Cryocoolers are popular for surface science experiments such as ARPES (angle-resolved photo emission spectroscopy), XPS (X-ray photoemission spectroscopy), AES (atomic emission spectroscopy), and other surface science techniques (Fig. 7.26).

7.6.8 Cryostat for Nano-Optics

This application was discussed in great detail in Sect. 7.5. It is the single most important area of research in the past few years. Scientists who are testing nano-

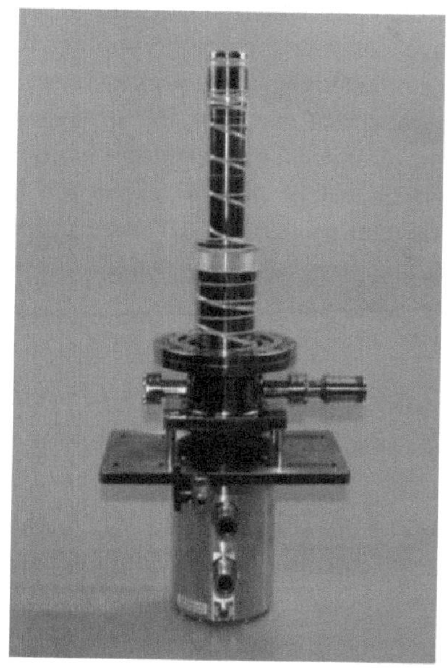

Fig. 7.25 The cryostat for ultra-low vibrations and ultra-high vacuum

Fig. 7.26 A cryocooler mounted on a UHV vacuum chamber for surface analysis. (Courtesy Prevac, Poland)

optics typically test a nanomaterial sample, such as a quantum dot or a 2D material, where the optical signal is weak, and there can be single-photon counts. This is done over a long period of time, from days to weeks. This necessitates holding the feature (few nanometers in size) steady in one place without vibrations and drift. Reducing the vibrations to the nanometer level when a cryocooler is inducing the vibrations can be a design challenge. A typical cryocooler has vibrations in the 10–15 μm at 2 or 2.4 Hz. These vibrations have to be isolated from the sample using novel techniques which are discussed in this chapter. An obvious challenge is that the need of strong thermal coupling between the sample and the cryocooler for low sample temperature fights against the requirement for low vibrations. The stronger the thermal coupling, the stronger the vibration transfer. The second challenge in the design is that materials with low thermal conductivity generally have high coefficient of thermal expansion (CTE), as these materials can also have poor vacuum properties. All these properties of the materials of construction and the design have to be considered.

7.6.9 XLV Cryostats for Atomic Microscopy (STM, AFM, SEM)

For microscopy there are two main requirements, ultra-low vibrations in the atomic scale and ultra-high vacuum for sample surface cleanliness. The Solvay cryocoolers have been coupled with a gas gap interface for extreme low vibrations (XLV)

Fig. 7.27 ULV cryostat use on an SEM microscope. (Courtesy Prof. Dan Rich, Ben Gurian University, Israel)

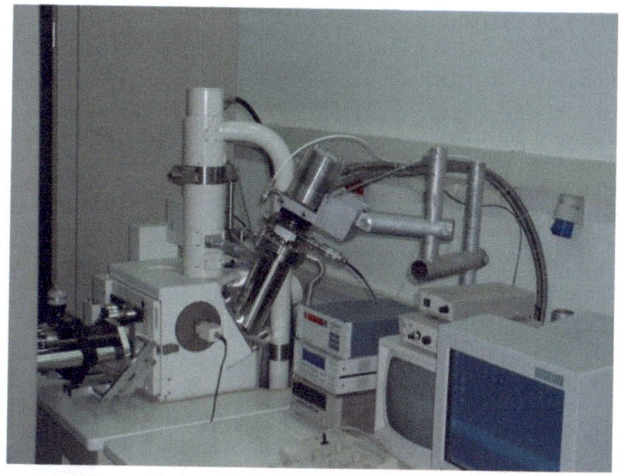

Fig. 7.28 ULV cryostat use on an STM. (Courtesy RHK, Inc.)

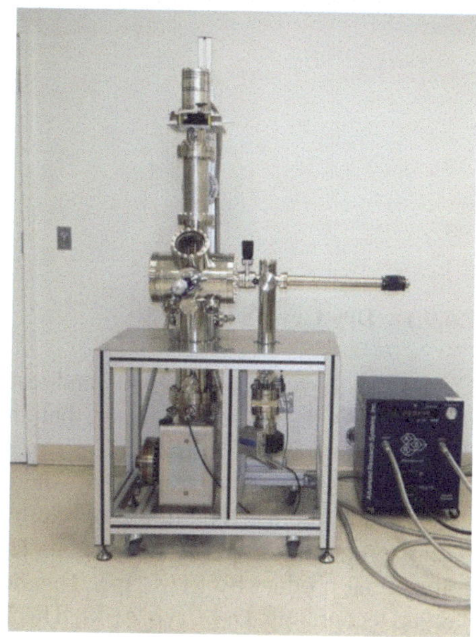

performance for mapping the atomic surface of materials of interest. The gas gap technology, discussed in detail in Sect. 7.5, has a long history of extreme low vibration applications.

A typical set up for an XLV atomic resolution experiment requires a cryocooler with a low-vibration design. This minimizes the induced vibrations to the XLV instrument such as an STM or AFM. Two examples are given (Figs. 7.27 and 7.28).

Fig. 7.29 A cryocooler mounted to an AFM for atomic imaging. Requires ULV and XHV design. (Courtesy Raschke Nano-Optics Group, University of Colorado, Boulder)

7.6.9.1 Dry Cryo-STM

Cryo-STMs have been developed and tested since the mid-1980s. However, since the early 2000s, it became clear that the use of liquid helium would prevent the adaptation of this instrument in several labs. The use of a cryocooler for an application as demanding as an STM where the spatial resolution of 1 Å is necessary was a real challenge. The first group to overcome this difficulty was Prof. George Nazins group at University of Oregon. The instrument was later developed into a commercial product by RHK, Inc. This demanding application also uses the ARS gas gap technology. Dry Cryo-AFM: This instrument was pioneered by the Raschke Nano-Optics Group, University of Colorado, Boulder (Fig. 7.29).

The Raschke group (all images and photos are courtesy Raschke Group) used the Solvay cryocooler with a gas gap in two configurations.

- Cryocooler supported on the table.
- Cryocooler supported off the optical table.

As the cryocooler is the component inducing the vibrations, the less mechanical influence it has on the optical table the better. For this reason, the Raschke group supports the cryocooler off the optical table.

Chapter 8
The Diaphragm Pressure Wave Generator

Alan Caughley

Abstract Diaphragm pressure wave generators have been combined with Stirling-type pulse tubes to produce industrially robust cryocoolers. Practical cooling powers have been demonstrated in the range of 10s of W up to 1500 W at 80 K, and the technology is ideally suited to liquefaction of air, nitrogen, oxygen and methane. The metal diaphragm of the pressure wave generator enables a variable volume compression space without clearance gap or oil seals. Hence, a clean cryocooler working gas can be maintained long term with an oil-lubricated reciprocating mechanism. Challenges with the imbalance of the high gas force on the diaphragms and efficiently creating a high-frequency short-stroke high-force reciprocating movement have been overcome. Cryocoolers have been made and demonstrated to economically produce liquid cryogens in an industrial setting. Diaphragm-based cryocoolers are now available commercially.

8.1 Introduction

The diaphragm pressure wave generator was developed in response to a need of high-temperature superconductor (HTS) applications in the power industry [1]. Cryogenically cooled transformers, cables and fault current limiters needed an industrial-styled cryocooler of a cooling capacity between the state-of-the-art lab or space-based cryocoolers and liquefaction plants. The industrial cryocoolers needed to produce from 100s of W of cooling to a few kW of cooling at 60–80 K, be long lived, efficient, robust and reliable, cheap and simple to maintain. Pulse tube refrigerators show much promise as a simple cold head with the potential for a very long life due to having no moving parts in the cold space. Pulse tubes however need to be driven by a pressure wave generator, a mechanical device that does the hard work in the cryocooler. Linear motor systems based on the Oxford pressure wave

A. Caughley (✉)
Mechanical Engineering, Callaghan Innovation, Christchurch, New Zealand
e-mail: Alan.Caughley@callaghaninnovation.govt.nz

M. D. Atrey (ed.), *Cryocoolers*, International Cryogenics Monograph Series,
https://doi.org/10.1007/978-3-030-11307-0_8

generator have proved successful at small scale, but rely on resonance, clearance gap pistons and flexure bearings—items that get difficult and expensive at large scales. What was needed was something simpler and more robust. Conventional piston–crank mechanisms are very simple and robust, produce good pressure waves and are cheap, but suffer from oil ingress past piston seals into the cryocooler gas space which limits lifespan. Hence the diaphragm pressure wave generator was developed [2]. The diaphragm pressure wave generator uses metal diaphragms to seal a conventionally lubricated motor/crank mechanism from the clean cryocooler working gas. The key innovation [3] was the use of a second diaphragm, rigidly connected to the working diaphragm, that balanced out the average gas pressure load so that the reciprocating mechanism only experienced the pressure wave. A second key innovation was the use of a hydraulic lifter mechanism to transform the short-stroke high-force movement of the diaphragm into a longer stroke lower force movement that could be efficiently produced by a motor and crank.

Since its invention, the diaphragm pressure wave generator-based pulse tube has shown utility in other applications than HTS [4, 5], primarily in on-site production of modest quantities of cryogens such as liquid nitrogen or oxygen. For liquefaction in remote areas, on-site production is better than trucking and storing liquid produced by distant liquefaction plants. Distillation columns work best over 10 kW of cooling power. If much less is needed, then the capital and running costs are too high to justify a full-scale liquefaction plant. An example is the test site for the DPWGs in Christchurch, New Zealand, where the local gas company has a distillation column, which is usually used for producing high-purity oxygen. The distillation column formerly produced liquid nitrogen for a small customer base, but the DPWG-driven cryocooler option can make the nitrogen for a third of the cost of the separation column so is used most of the time. Other users are remote users who are a significant distance from the nearest liquefaction plant, such as for biological storage, where security of supply and cost of shipping make economic sense to generate LN2 on site. Medical centres in non-industrial areas can use cryocoolers to liquefy oxygen and store for use. The lowering cost of solar power and short start-up times offered by DPWG cryocoolers makes generation of oxygen during daylight for availability later feasible.

8.2 Working of the DPWG

The DPWG, shown in Fig. 8.1, utilizes two diaphragms, one to provide the pressure wave in the compression space (the driving diaphragm). The gas spring has a large closed volume and an identical diaphragm with the same displaced volume as the driving diaphragm. The resultant pressure wave in the gas spring is small. The two diaphragms are oriented axially and are rigidly connected together by a reciprocating ‘drive piston’. The diaphragms at each end of the drive piston mean that it is very stiff radially, but free to move axially with the motion which is very much like the flexure bearing system often found in linear motor pressure wave

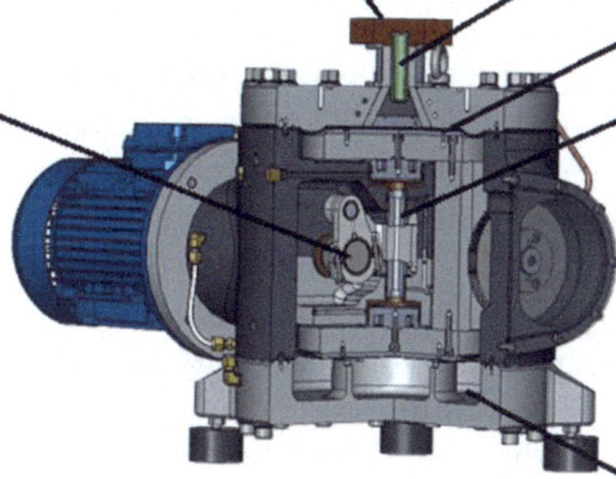

Fig. 8.1 Cutaway view of a diaphragm pressure wave generator and axial pulse tube cryocooler

generators. The gas spring diaphragm then effectively balances the average force on the driving diaphragm leaving the crank mechanism to experience only the pressure wave forces.

An electric motor rotates a crankshaft whose con rod reciprocates the drive piston. Because the diaphragm stroke is very short, and load high, a gearing system is incorporated between the crank and the drive piston. In early versions, this was in the form of a lever, but all models now use a hydraulic lifter with small master piston and large slave piston. The hydraulic lifter allows a higher gearing ratio in a smaller space and has quieter and smoother operation than the lever.

Twin crankshafts mean that all the vibrations from the reciprocating components happen along the central axis of the machine which is simply balanced at all speeds by four counterweights, symmetrically placed on each shaft. A pair of gears maintains the timing of the shafts and provides enough flywheel action to absorb the expansion energy, returning it to the next stroke's compression.

The diaphragms seal the working gas from the drive mechanism. The shape of the diaphragms has been carefully optimized to balance the opposing shapes required for maximum flexure and holding pressure whilst remaining under the fatigue limit for the material. The diaphragms act as flexure bearings being very stiff radially and free to move axially. Like flexure bearings, the theoretical life of the diaphragms is infinite when run below the fatigue limit of the material. The diaphragm has the advantage over the flexure bearing in that it has no cut-outs to produce edge stress concentrations. However, the stiffness of the metal membrane limits the usable stroke to around 1% of the outside diameter of the diaphragm.

8.2.1 Integrating Pulse Tubes with the Diaphragm Pressure Wave Generator

The pulse tube cryocooler consists of seven parts: a pressure wave generator (in this case the DPWG), an aftercooler to remove heat at ambient temperatures, a regenerator, a cold heat exchanger to absorb heat at cryogenic temperatures, the pulse tube, a warm heat exchanger and a phase shifter (commonly an inertance tube and reservoir for larger pulse tubes).

For DWPG-driven pulse tubes, best results are achieved with the pulse tube as closely coupled to the DPWG as possible. Fortunately, the flat face of the diaphragm and its corresponding housing lend well to incorporating a slotted heat exchanger-type aftercooler and pulse tube without any interconnecting ducts.

8.2.2 Coaxial and In-Line Pulse Tubes

Coaxial pulse tubes, such as in Fig. 8.2, present a cold surface at the end of a short cold finger. Since the driving piston can be either the top or bottom diaphragm, then the cold finger can be inserted down the neck of a Dewar to condense directly into the Dewar. A condensing surface needs to be mounted on the cold face to present enough area to the gas to condense. This method was used successfully on a liquefier with the 100 W cryocooler, producing a liquefier with approximately 20 L/day liquid nitrogen production. The arrangement was very effective but the vibration from the pressure wave generator—onto which the Dewar was directly mounted—propagated through the aluminium Dewar shell and was very noisy. Additionally,

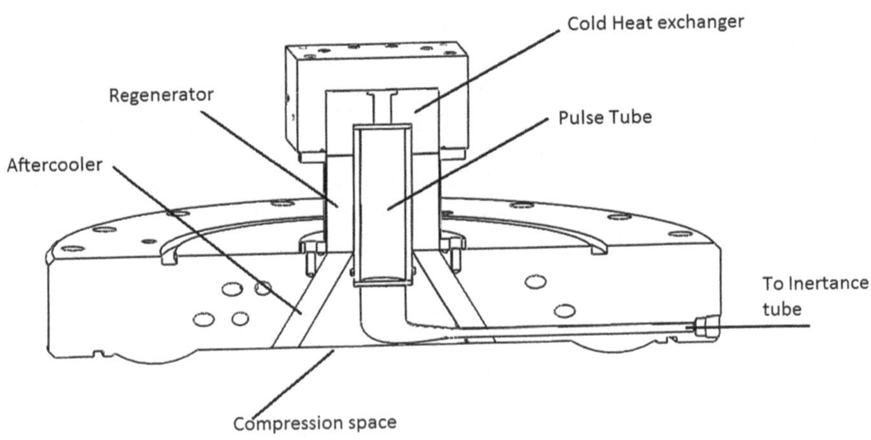

Fig. 8.2 Cross section of the PT10090 coaxial pulse tube [4]

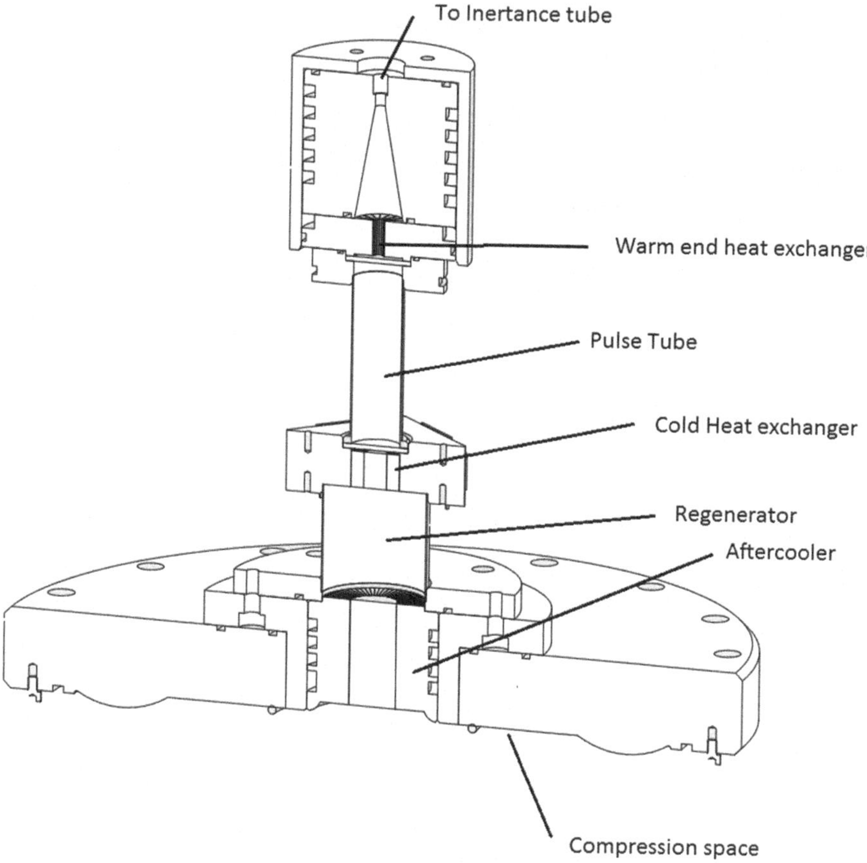

Fig. 8.3 The in-line pulse tube mounted on the dWPG top plate [6]

the short length of the coaxial pulse tube meant that the cold surface was still in the neck of the Dewar, which was not so well insulated, and resulted in some heat loss.

The **in-line pulse tube** is a better and more effective arrangement. Figure 8.3 shows an in-line pulse tube mounted on the compression side of a DPWG.

A heat exchanger is used to condense the gas which drains into the storage Dewar via an insulated pipe. The in-line arrangement has a cold surface in the middle of the pulse tube and warm gas is fed to the pulse tube heat exchanger via a copper tube. The tube wound around the cold heat exchanger cools the gas down, which eventually liquefies. Careful choice of the tube diameter and length is necessary to make sure that liquid does not build up and block the flow, causing a reduction in the temperature of the pulse tube leading to freezing and blocking of the heat exchanger. The liquid drains into the Dewar via a vacuum jacketed transfer tube.

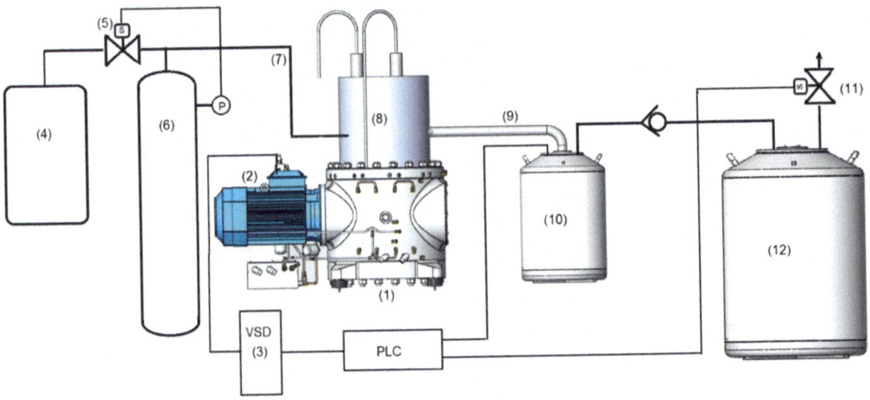

Fig. 8.4 Schematic of a liquefier system with the PTC1000 DPWG cryocooler

8.3 Integrating the Cryocooler into a Liquefaction System

Figure 8.4 shows a DPWG cryocooler integrated into a liquefier.

Nitrogen gas is generated using a pressure swing nitrogen separation system (4); a buffer tank (6) and valve (5) are used to maintain a constant inlet pressure. The warm nitrogen gas flows into the system via a standard gas tube (7). The pulse tubes are inside the cryocooler's cryostat (8) where the warm gas is cooled and then condensed. The liquid then flows via a vacuum-insulated line (9) into a buffer Dewar (10). The DPWG (1) turns on and off via a variable speed drive (3), controlling the motor (2). When the buffer Dewar is full, or product is desired, the PLC controller operates a relief valve (11) on the storage Dewar (12) vent. When the storage Dewar pressure drops below the pressure in the buffer Dewar, liquid will flow past the check valve the storage Dewar.

8.4 Use of the Device

The diaphragm pressure wave generator, at the time of writing, is still a new technology with beta level units in use at various industrial sites.

The beta testing site for the DPWG based cryocoolers was Southern Gas Services in Christchurch, New Zealand. Southern Gas supplied local industry with industrial process gases and had a small customer base for liquid nitrogen. New Zealand's liquid nitrogen supply was dominated by air separation plants in conjunction with the steel mill at Glenbrook near Auckland in the North. Hence liquid nitrogen for the South Island was shipped and trucked over 1000 km from source to use. The price of liquid nitrogen was NZ $10 per litre. Southern Gas services ran a small distillation column for producing high-purity oxygen. The column could be used to produce

Fig. 8.5 The first PTC1000 cryocooler in operation at SGS in 2015

liquid nitrogen but quantities were much in excess of that required by Southern Gases customer base, so production of liquid nitrogen was relatively expensive. The PTC1000 cryocooler can produce 11 L/h of LN2 which matches SGS's customer usage, and the DPWG's continuous production has enabled SGS to secure new customers. The first PTC1000 cryocooler was installed in 2015 and was run for a year [7] before being replaced by a beta prototype. The test site has provided valuable running experience for developing the DPWGs (Fig. 8.5).

References

1. Sheahen ETP, Mcconnell B (2001) Cryogenic roadmap. US Department of Energy Superconductivity Program for Electric Systems
2. Caughley AJ, Haywood DJ, Wang C (2008) A low cost pressure wave generator using diaphragms. In: AIP Conference Proceedings, vol 985
3. Caughley AJ, Haywood DJ (2015) Cryogenic refrigerator system with pressure wave generator, US 8984898
4. Emery N, Caughley A, Glasson N, Meier J (2012) Co-axial pulse tube for oxygen liquifaction. Adv Cryog Eng 57A:183–189
5. Caughley A et al (2016) Commercial pulse tube Cryocoolers producing 330 W and 1000 W at 77 K for liquefaction. IEEE Trans Appl Supercond 26(3)
6. Emery N, Caughley A, Glasson N, Tucker A, Gschwendtner M (2011) Development of a high frequency pulse tube. In: Cryocoolers, vol 16. ICC Press, Boulder, pp 175–182
7. Caughley A et al (2016) Results from a year of running a 12 L per hour pulse tube liquefier in an industrial application. In: Cryocoolers, vol 19. ICC Press, Boulder, pp 623–630

Chapter 9
Cryocoolers for Healthcare

Wolfgang Stautner

Abstract This chapter shows the great impact cryocoolers made and continue to make in their emerging role as workhorse in the healthcare industry. Their great impact is particularly noticeable in MRI, PET, and many other small or large healthcare applications; for example, metabolic imaging and preclinical applications that would be unthinkable without the use of cryocoolers. The sustainability effect and carbon footprint reduction of cryocoolers, e.g., on MRI systems due to their ability to recondense helium and other refrigerants cannot be underestimated. The continuing increase in cryocooler cooling power capacity, their industrial robustness, and their ability to deeply embed pulse tube coolers in a cryostat or magnet system with very long service intervals continue to improve their usability for "wet" and "dry" healthcare applications.

9.1 Introduction

9.1.1 The Importance of Cryocoolers for the Medical Industry: Saving Helium

The world's helium resources remain limited despite recent findings and current oversupply. This is because helium is wasted once released to the atmosphere.

This is a fact known to engineers and scientists working in cryogenics or in relevant industries as shown in Fig. 9.1. Scientists around the world continue to experience challenges with respect to cost and availability of helium for their equipment and long-term budget planning.

As a result, GE has made great strides on material sustainability in MRI magnet manufacture following mottos "We c green" and implementing "green

W. Stautner (✉)
GE Research, Biology and Applied Physics, Cryogenics & Superconducting Magnet Applications, One Research Circle, Niskayuna, NY, USA
e-mail: stautner@ge.com

M. D. Atrey (ed.), *Cryocoolers*, International Cryogenics Monograph Series,
https://doi.org/10.1007/978-3-030-11307-0_9

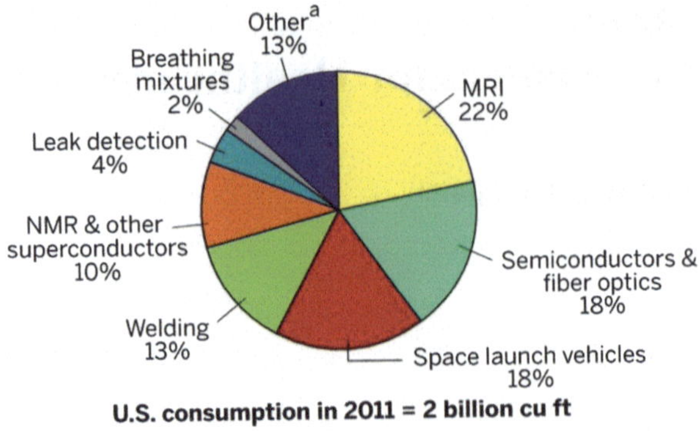

Fig. 9.1 US helium consumption in 2011 [1]. Reproduced with permission from C& EN (Reisch [1]). Copyright 2013 ACS

technologies." Furthermore, GE has made a long-standing commitment to ensure helium is conserved for years to come for the future.

Figure 9.1 not only shows the impact the MRI and NMR industry has on helium consumption but also indicates the potential of significantly reducing this cost. The efforts of the MRI industry to increase equipment and manufacturing efficiency and to recycle helium have now been very well noticed by the gas industry [2].

That said, even more helium will be required for other applications appearing on the horizon, e.g., for filling large helium airships (e.g., Airlander 10) at a cost of currently $330 k$ per fill [3].

The superconducting coils of a persistently operated MRI magnet system are traditionally helium bath cooled. Currently, over 3000 MRI units are produced worldwide annually adding to an installed base of 35,000 units with fill volumes varying from 1000 to 2500 L, depending on size and field strength.

According to GE Chief Magnet Manager Peter Jarvis, 5 million liters are roughly used in MRI magnet production and approximately 6 million liters for maintaining devices already installed in hospitals and clinics. Every new installation in a hospital requires infusions of a few hundred liters to compensate the helium lost in transition to the customer site [4]. Tom Rauch, the GE Healthcare Global Sourcing Manager mentions: "Depending on the age, model and location of the machine, MRIs need to be topped off with helium at least once per year—not unlike topping off the coolant in your car. To do this, GE contracts with helium retailers to deliver helium and related service to customers" [5].

9.1.2 Saving Helium at the Factory

To reduce general helium consumption several points need to be addressed at the manufacturing level in the factory already, e.g., optimized helium heat transfer based on simulated computer models to minimize helium use for magnet precool, to optimize helium transfer lines, for investment in helium liquefiers and capture of helium gas in high pressure gas bottles.

9.1.3 The Role of Cryocoolers for MRI and NMR Systems for Saving Helium

Liquid helium reduction for MRI magnet systems has come a long way which would not have been possible without the use of cryocoolers. Table 9.1 below gives an outline of those efforts and technology steps taken.

Historically, MRI systems of the early 1980s in the absence of cryocoolers, boiled away 0.6–1 L/h of liquid helium (equivalent to >10,000 L of helium gas/day) of a liquid reservoir volume of approximately 2000 L that was not being recovered.

After arrival of the shield cryocoolers this boil off was then reduced to 35–60 mL/h followed by GE's novel recondensing effort in the late 1990s when no helium was lost to environment anymore. However, liquid helium volume for precool, initial fill, and associated cost of energy **was substantial**.

9.1.4 Saving Helium When Decommissioning a System

A further saving of helium is possible if liquid helium is transferred, back to liquid helium Dewars, rather than quenching the magnet at the hospital, and releasing helium back to environment. The current inventory of the installed MRI base, depending on fill levels, stores 35–70 million liters of liquid helium. This seems like a small figure but with helium prices of 3–15$ on average/liter for some users, saving liquid helium may be an attractive option.

Figure 9.2 shows the latest, conventional 3 T GE MRI scanner as of 2015 running with 1 GM type cooler. The most recent brain imaging applications giving new insights into the working structure of the brain (see Fig. 9.3). These insights had not been possible earlier.

Table 9.1 Helium saving efforts in MRI industry (past and future), see also [6]

Helium use reduction in MRI industry—paradigm changes		Helium
Cooldown	Using nitrogen and helium	Fill
Cooldown reduction	Use helium and recovery system	Fill/recover
Boil off reduction	Dual-stage shield cryocoolers	Refill
Boil off elimination	4 K cooler	No refill
Reservoir reduction	Use minimum of liquid helium	Gas filled
Reservoir elimination	Conduction-cooled, cryocoolers	Zero

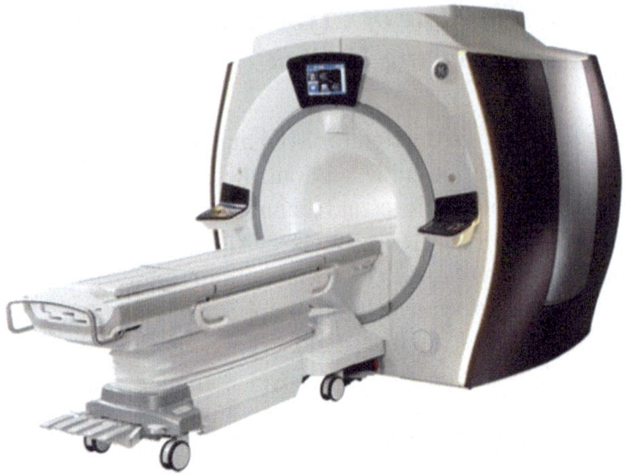

Fig. 9.2 GE MRI 3 T scanner

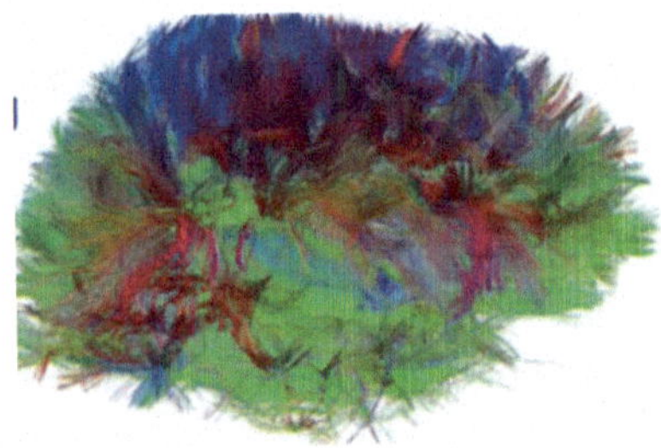

Fig. 9.3 MRI image of brain connections and active regions

9.2 Conventional MRI Scanner Schematics and Cryogenics

We will now take covers off the MRI system as shown in Fig. 9.2 and vertically cut through it. Figure 9.4 exemplary shows the main components of an MRI magnet system with its plurality of superconducting (SC) magnet main and bucking coils and associated shim and moving metal coils in a liquid helium filled vessel. The helium vessel is protected by a thermal shield against thermal radiation heat influx from the outer vacuum vessel walls. A vertical tube penetration allows for direct access to the helium bath for ramping the magnet with current leads and for routing other instrumentation wires.

A second penetration through the vacuum space is required for accommodating the dual-stage cryocooler with its attached recondensing cup for liquefying boiling off helium gas.

High performance gradient systems which are pulsed intermittently superimpose linearly varying fields on the uniform background field to vary the resonant frequency within the field of view (FOV). The configuration of other magnet design related topics can be found in greater detail in [6].

The inner bore region of an MRI system highlights challenges the cryogenic engineer has to cope with given the distance from the inner vacuum vessel tube

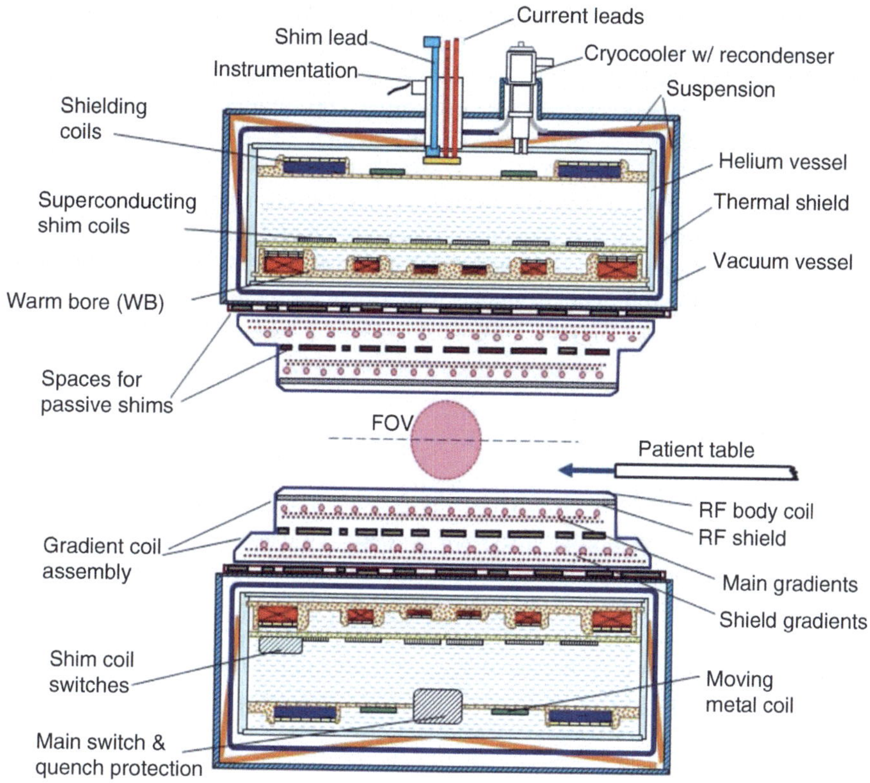

Fig. 9.4 Generic layout of generic MRI magnet system [6, 7]

(warm bore) to the first cryogenic temperature is usually only 25–30 mm. Other design constraints derive from the choice of the cryocooler used in a particular design.

Most mainstream MRI magnets work with one Sumitomo RDK 412 dual-stage cryocooler delivering a cooling power of 1.2 W at 4 K. The engineer therefore has to design for a low enough background heat load that allows for sufficient assembly tolerances and long coldhead maintenance intervals.

9.3 Cryocooler Integration

Since their commercial availability in the 1980s, cryocoolers have become excellent and versatile tools for a diverse range of cryogenic applications, from heat sinking a cold mass, to distributed cooling and even for liquefying or dedicated component cooling. At the cryogenic design stage, the cryogenic engineer needs to know in which way the chosen cooler, Gifford-McMahon (GM), Pulse Tube Refrigerator (PTR) or Stirling type, can be integrated in a cryogenic structure and in particular how the thermal links to the cryocooler cold stages should be made.

Generally, the cryocooler interface cost is driven up significantly when demanding minimum temperature difference and by associated requirements, e.g., low vibration transmission, or when it is necessary to retract or remove the cooler. Cryogenic designs have to balance this contact quality effort and make a trade-off against what is needed and is feasible for assembly and cooler maintenance. The most common design solutions for dedicated applications are therefore listed below and illustrated in the following examples [8].

9.3.1 *Options Overview*

Cryocoolers for thermal mass cooling usually can be:

- Fixed bolt-on, braided, internal cryocooler parts removable
- Sleeved sleeve bolt-on or braided, cryocooler removable from sleeve/serviceable
- Un-sleeved cryocooler exposed to cryogenic atmosphere, e.g., GM/PTR cooler in helium gas
- Retractable temporarily engage/disengage from cold mass, e.g., after initial cooldown, sleeved

9.4 Cryocooler Integration Examples

A cryocooler can also be integrated but made detachable from its contact stages for a number of reasons. For example, in case of a sudden, unplanned power outage and following compressor shut down resulting in a loss of cooling power, a thermal short would be created that leads to high heat loads to the cooling components (e.g., thermal shield, current leads, cold mass). For large thermal mass cooling (e.g., during cooldown, see Sect. 9.5) a shielded drive mechanism can be installed. The drive mechanism moves the cooler cold end to the cold mass interface, maintaining a specified contact pressure that compensates for the cold mass shrinkage. To avoid a permanent parasitic heat load on the cold mass that can be substantial for larger coolers, the cooler is then retracted automatically [9].

Figure 9.5 shows MRI applications using high-temperature superconductors (HTS) and cooling approaches as discussed in this section.

The cryocooler for the so-called open C-magnet (1) is rigid and sleeveless, bolted onto a cold plate. From the cold plate, braids are routed to the copper sheets that are sandwiched between the individual pancake coils. (2) and (3) show various solutions for extremity scanners for different superconducting wires using thermosiphon technology and cryogen liquefaction with zero boil off.

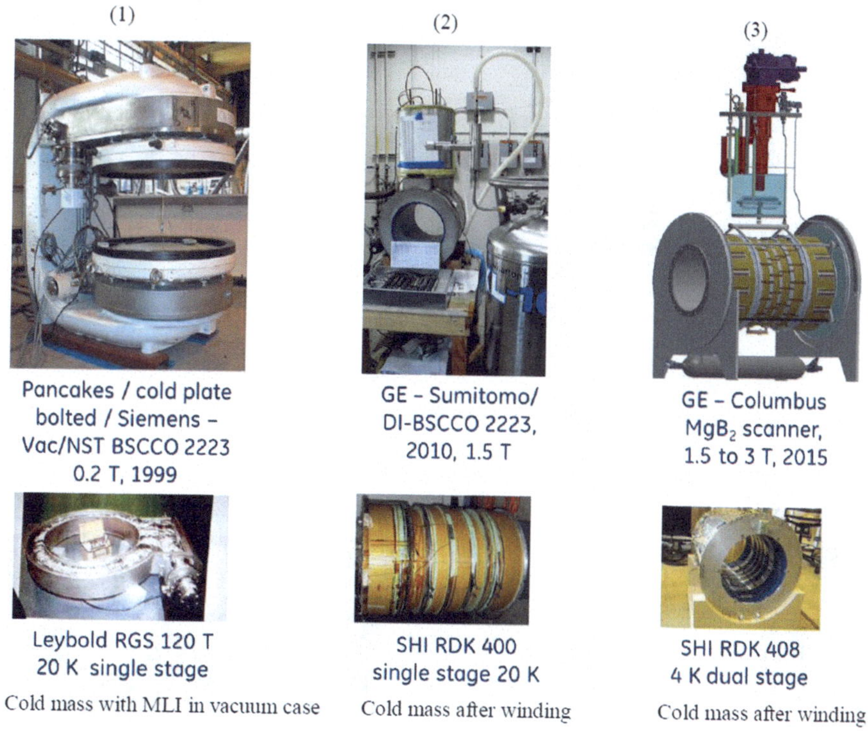

Fig. 9.5 Cryocooler integration—small-scale HTS MRI

1. Cooling technique: conduction cooling, cryogen: n/a, cooler: single-stage GM [10]
2. Cooling technique: thermosiphon, cryogen: neon, cooler: single-stage GM [11, 12]
3. Cooling technique: thermosiphon, cryogens: helium or hydrogen, cooler: dual-stage GM [13]

A good example for directly embedding a pulse tube cooler (PTR) in the coil former of the superconducting magnet cold mass itself is the head imaging prototype system built by GE Global Research back in 1995 (see Fig. 9.6).

Technically, this is a bold approach since the cooler cannot be removed. However, due to the absence of a moving piston no vibrations are transmitted to the superconducting magnet structure. Furthermore, this cooler does not need to be serviced but can be purged of contaminants when warmed up to room temperature, if necessary.

Direct cryocooler cold mass bolt-on is also an interesting option for a number of small-scale MRI magnet systems using medium or high-temperature superconductors (HTS), see also (1) in Fig. 9.5. Another early, interventional helium-free LTS MRI magnet system [6] (see Fig. 9.7) built by GE used fixed installed cryocoolers.

Fig. 9.6 GE 0.5 T conduction-cooled head imaging system [6]

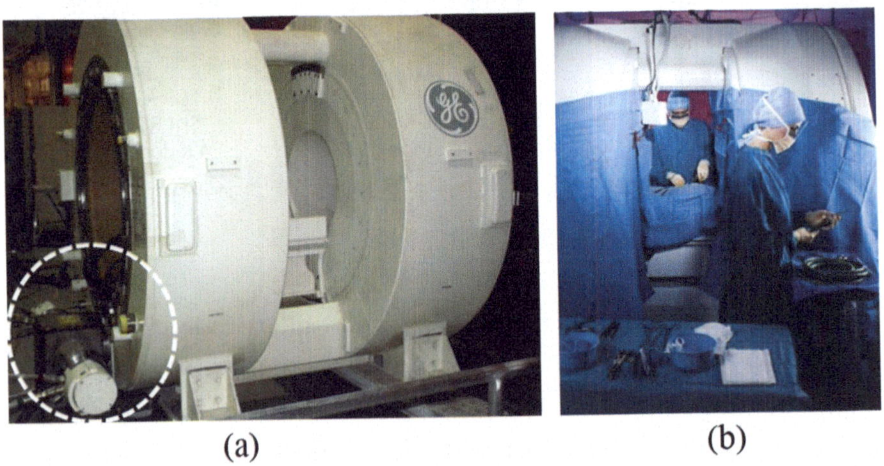

Fig. 9.7 (**a**) Cryocoolers (2) (dotted line) in cold box and in contact with SC coils via heat bus (**b**) Surgeons in the operating theater

Both coolers shown in Fig. 9.7 are sleeved to allow the removal of the cooler from the cold plate during service. The cold plate or heat bus extends from the left to the right of both Helmholtz style superconducting Nb_3Sn coils, operating at 10–11 K.

Note that un-sleeved GM coolers can also be serviced if required, by carefully removing the internal piston assembly from the cooler shell tubing. In this case, a warm swap to room temperature may be required, which requires a cold mass warm up, unless the cooler envelope can be heated to a high enough temperature to allow servicing or refurbishments.

For more traditional MRI systems, the cooler is either non-sleeved and directly exposed to helium gas atmosphere or vacuum sleeved but independent of the cryostat vacuum.

9.5 Cryocooler for Cryostat and Magnet Cooldown

Initial superconducting magnet cooldown is one of the many operating parameters that the engineer needs to design for. Although the steady state cryocooler performance is known, thanks to the published load map of the cooler manufacturer, the cooldown time can only be estimated properly, if targeted background heat load of cryostat and cooler performance from room temperature down to operating temperature is known (Fig. 9.8).

The diagram shows the performance of the standard 1.5 W at 4 K SHI RDK 415 cooler [courtesy of John Pfotenhauer of Univ. Madison and Ryan Taylor of NIST, Boulder].

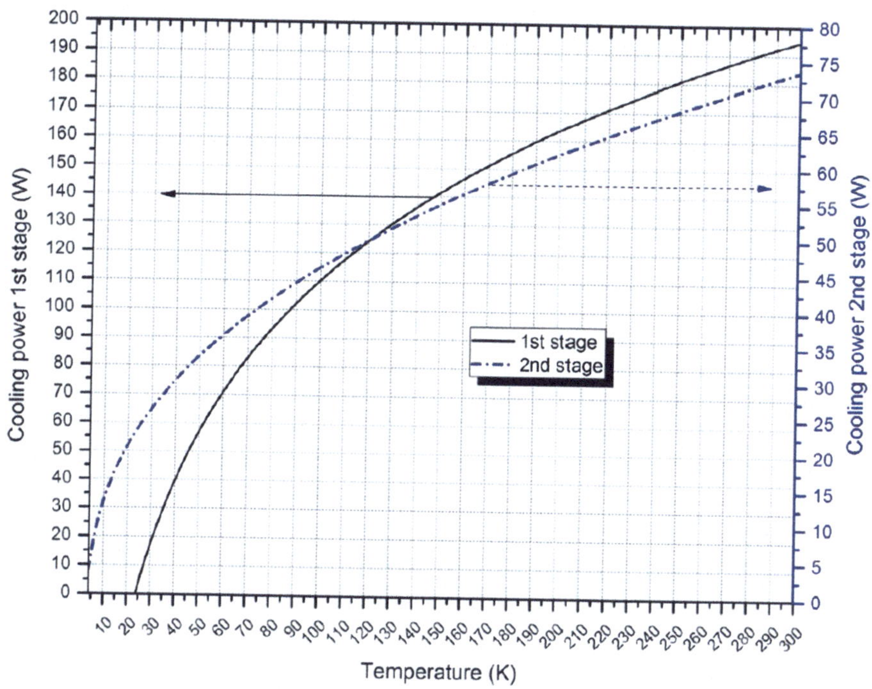

Fig. 9.8 Cryocooler performance diagram for operating temperature and fit for cooldown from room temperature (Pfotenhauer J, private communication)

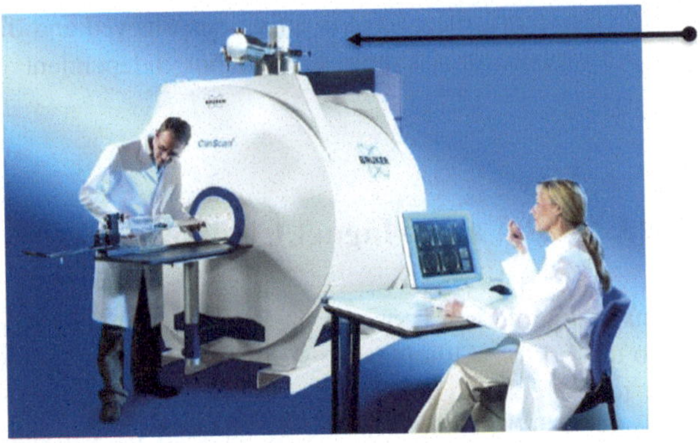

Fig. 9.9 ClinScan, B0 = 7 T, bore diameter = 30 cm, with active shielding [6]

9.6 Cryocoolers for Preclinical Systems

Scanners used for pharmaceutical and biomedical research on animals can take advantage of a much smaller bore and siting footprint. Bruker BioSpin is the leading manufacturer in this area. Figure 9.9 shows the ClinScan system used in animal MRI/MRS metabolism studies.

Since the piston of GM coolers transmits vibrations to the inner cryostructure with each cycle as mentioned, high-field magnet systems, especially NMR high-field magnet systems are equipped with pulse tube coolers, mainly from manufacturer Cryomech.

9.7 Cryocoolers for Standalone Metabolic Imaging Devices

The GE Spinlab system [6] is a so-called pre-polarizer, basically an MRI add-on device, that holds significant promise in metabolic diagnostics. It uses dynamic nuclear polarization (DNP), particularly, hyperpolarization of 13C, as a tool for a wide range of applications: study of metabolic processes for disease diagnosis, determination of tissue pathology and organ viability, angiography and perfusion studies, interventional MR, catheter tracking, and chemical ablation therapy.

With 13C hyperpolarization, metabolic processes within the body are made visible by injecting a solution into the human body that has been pre-polarized by cooling a vial filled with pyruvate to 0.8 K and polarizing it at, e.g., 3.35–5 T or higher prior to the imaging process. The injection sample has to be cooled down to satisfy the procedure time window, since its relaxation time has strong temperature dependence (Fig. 9.12). This method provides highly sensitive process for prostate cancer detection and many other applications.

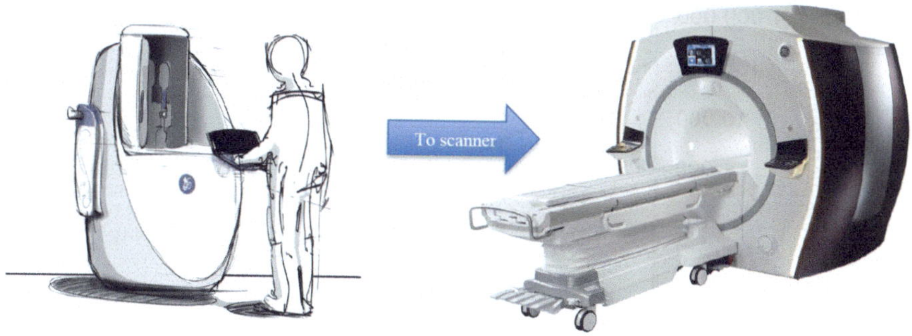

Fig. 9.10 Artist's rendition of ^{13}C polarizer next to MRI scanner

Fig. 9.11 (**a**) 3–5 T magnet and cryostat (sub-atmospheric operation); (**b**) completed automated MRI stand-by unit

As an add-on to an MRI system, Spinlab systems, based on either 13C or 128Xe will be used. Xenon hyperpolarization would be mainly applicable for low proton density organs, such as lungs, to provide an excellent signal enhancement. Figure 9.10 shows an artist's sketch of the fully automated Spinlab built by GE (Figs. 9.11 and 9.12).

In November 2010, GE Healthcare together with UCSF introduced this technology for the first time on a human patient, to conduct real-time metabolic imaging, after prior tests on animals.

This new approach improves assessing the precise outlines of a tumor, its response to treatment, and how quickly it is growing.

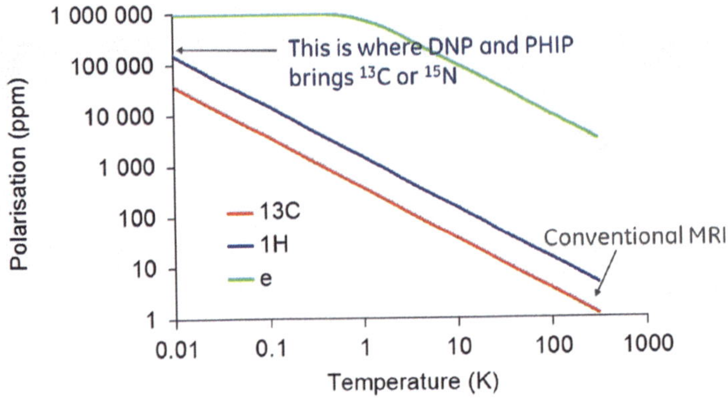

Fig. 9.12 Increasing the signal to noise ratio with high polarization, relaxation time of polarized ^{13}C pyruvate vs temperature [6]

9.7.1 Description of Cryosystem

The Spinlab system consists of a 3.35 T or a 5 T superconducting magnet fitted in a cryostat. Magnet and thermal shield are cooled using one RDK 415 Sumitomo cooler.

The thermal shield is in contact with the first stage of the cryocooler using copper braids. Magnet cooling is accomplished by a recondenser that is conductively linked to the cryocooler heat bus and allows the magnet to run sub-atmospheric. The cryogenic insert is located within the thermal shield. Part of the cooling power of the GM cooler is absorbed by the insert to maintain the function of a closed cycle sorption pump. The main elements, helium vessel, sorption pump, main helium condenser, and thermal bus with attached thermal switch, are shown in Fig. 9.13.

In this unique configuration, one cryocooler serves a double, segmented heat bus with different, time dependent operating characteristics.

Figure 9.14 shows the top plate of the vacuum can in a view from above, with the cryocooler and the suspension elements and heat bus.

9.8 Conclusions

Without the use of cryocoolers, MRI/NMR magnet systems would not be commercially feasible in a routine hospital environment. The invention of the cryocooler allowed the redesign of MRI cryostats that in the past, boiled off liters of helium per hour.

In fact, the reduced cooling power of a cryocooler operating at 4 K forces the designer to minimize and optimize the thermal budget to the extreme for zero

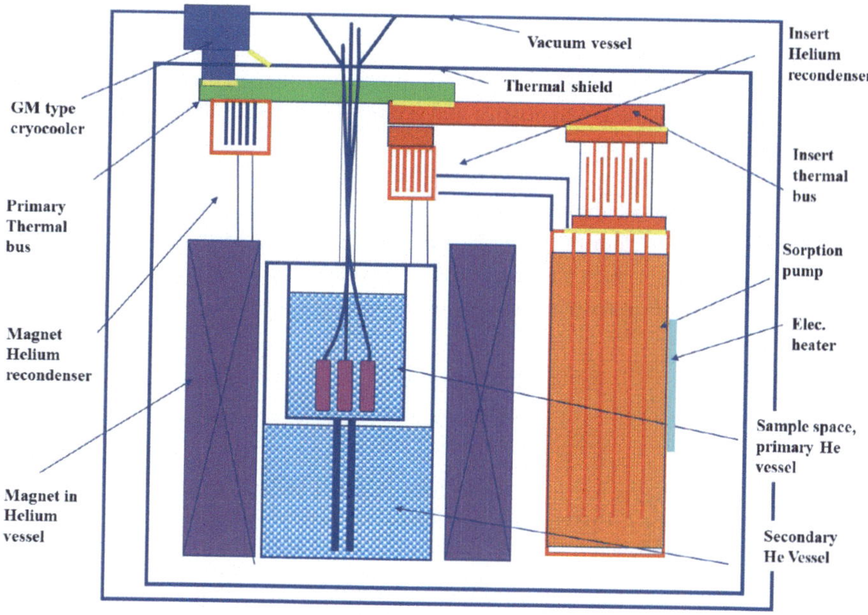

Fig. 9.13 Spinlab Cryogenic system schematics [14]

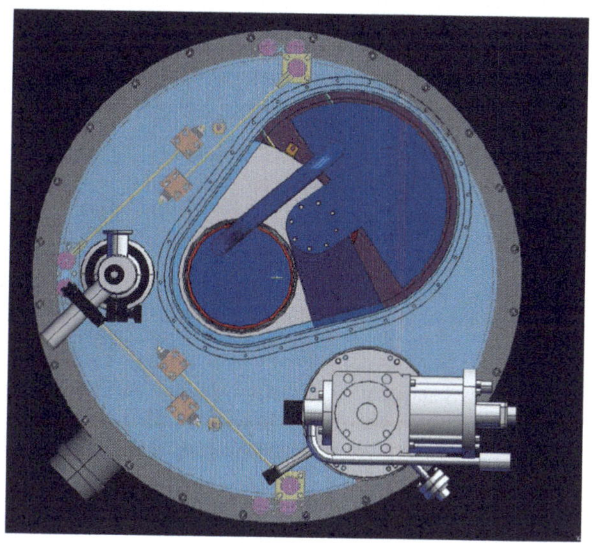

Fig. 9.14 Spinlab Cryogenic system schematics—**view from above**

helium boil off (ZBO). This is achieved by inventing new cryogenic techniques that sometimes take many years to develop.
On the other hand, the engineer needs to minimize cost to keep in mind that MRI, as the most important medical diagnostics and screening tool, should remain affordable for all.

The best design needs to be found between these extremes.

The road to future is clearly marked toward a helium-free environment that runs and controls itself by remote maintenance without direct user intervention. The cryocooler continues to play the major role in this scenario. Higher cooling powers with increased MTBF at reasonable cost remains the main goal.

References

1. Reisch MS (2013) Heavy demand NMR and other. Chem Eng News 91:19
2. Garvey M (2016) The 2016 worldwide helium market. Gasworld 54(9):50–55
3. O'Brien SA (2016) Helium airships are on the rise. Gasworld 54(9):40–43
4. Jarvis P (2012) GE sets sights on MRI helium conservation. dotmed.com
5. Rauch T (2012) The Helium Stewardship Act of 2012, Testimony before United States Senate Committee on Energy and Natural Resources
6. Lvovsky Y, Stautner W (2013) Novel technologies and configurations of superconducting magnets for MRI. Superconductor Sci Technol Top Rev 2013093001
7. Stautner W (2016) Review on superconducting technology of MRI scanners. In: 26th international cryogenic engineering conference, international cryogenics material conference 2016, Plenary talk 6, New Delhi, 2016
8. Stautner W (2016) Special topics in cryostat design. In: Weisend JG (II) (ed) Cryostat design, international cryogenics monograph series, Chapter 7. Springer, New York, pp 195–216
9. Stautner W (2012) Remote actuated cryocooler for SC generator and method of assembly the same. US20140100113A1
10. Davies FJ, Stautner W, Byrne AF, Wilson M (1999) An HTS magnet for whole-body MRI. In: Proceedings of the EUCAS'99, pp 1215–18
11. Stautner W, Amm K, Laskaris ET, Xu M (2007) A new cooling technology for the cooling of HTS magnets. IEEE Trans Appl Supercond 17:2200–2203
12. Stautner W, Xu M, Laskaris ET, Conte G, Thompson PS, van Epps C, Amm K (2011) The cryogenics of a thermosiphon-cooled HTS MRI magnet—assembly and component testing. IEEE Trans Appl Supercond 21:2096–2098
13. Stautner W, Xu M, Mine S, Amm K (2014) Hydrogen cooling options for MgB_2-based superconducting systems. AIP Conf Proc 1573:82–90
14. Urbahn J, Stautner W et al (2009) A closed cycle helium sorption pump system and its use in making hyperpolarized compounds for MRI imaging. In: ICEC Proceedings of the 22nd international cryogenic engineering conference, vol 22, pp 589–594

Chapter 10
Cryocoolers for Helium Liquefaction

Stefano Spagna, Javier Sesé, Miguel Gabal, and Conrado Rillo

Abstract In this chapter, we describe a gas liquefaction system and method, based on closed-cycle refrigerators, that utilizes the thermodynamic properties of gaseous elements to extract the maximum cooling power from the refrigerators. The system operates at elevated pressure, near and above the critical pressure, and with the gas in direct contact with the cryo-refrigerator body along its entire length. We show that this new liquefaction method dramatically enhances the liquefaction rate over previously known techniques. In this description, a cryo-refrigerator with N-number of stages operates in the neck of a thermally isolated container or Dewar. The first stage is the warmest and operates higher in the neck than the second stage that operates lower in the neck and below the first stage, and so on. Prior to liquefaction, the gas is cleansed to extremely high purity levels. In the case of helium, this is achieved in two steps. First, the higher vapor pressure impurities (i.e., air components like N_2, O_2, CO_2, etc.) are condensed in a closed-cycle refrigerator-based purifier. Then, the lower vapor pressure impurities (mainly H_2), which cannot be condensed in the cryogenic purifier, are eliminated by chemical adsorption in a specific getter material. The resulting high purity gas enters at the top of the neck of the liquefaction Dewar and is precooled by the upper first stage of the cryo-refrigerator, further cooled by the next stages, and then condensed by

S. Spagna (✉)
Quantum Design Inc., San Diego, CA, USA
e-mail: stefano@qdusa.com

J. Sesé
Departamento de Física de la Materia Condensada, Universidad de Zaragoza, Zaragoza, Spain
Instituto de Nanociencia de Aragón (INA), Universidad de Zaragoza, Zaragoza, Spain

M. Gabal
Departamento de Física de la Materia Condensada, Universidad de Zaragoza, Zaragoza, Spain
Servicio General de Apoyo a la Investigación-SAI, Universidad de Zaragoza, Zaragoza, Spain

C. Rillo
Departamento de Física de la Materia Condensada, Universidad de Zaragoza, Zaragoza, Spain
Instituto de Ciencia de Materiales de Aragón (ICMA), Universidad de Zaragoza-CSIC, Zaragoza, Spain

M. D. Atrey (ed.), *Cryocoolers*, International Cryogenics Monograph Series,
https://doi.org/10.1007/978-3-030-11307-0_10

the lowest stage of the cryo-refrigerator. Using these cryocooler-based techniques, helium recovery plants for applications requiring gas or liquid with extreme purity, and, liquefaction rates adapted to consumption, are now available everywhere. The combinations of analytical and numerical calculations nicely reproduce all the features of the liquefaction process.

10.1 Introduction

Closed-cycle refrigerators have benefited from gradual performance improvements during the last five decades [6, 11, 12, 15, 16, 25, 32, 33]. Nowadays, very reliable two-stage commercial Gifford-Mac Mahon (GM) and Pulse-Tube (PT) based cryo-refrigerator systems are available [5, 27, 28]. These systems comprise of a dc flow helium compressor that supplies through a rotary valve, an ac helium gas flow to the cryo-refrigerator. Thanks to rare earth-based regenerators, a nominal cooling power of 1.5–1.8 W at 4.2 K can be achieved at the second stage heat station of the cryo-refrigerator.

Initially developed to operate in vacuum to cool down devices by solid conduction cooling through thermal contact with the cryo-refrigerator cooling stages, researchers were employing closed-cycle refrigerators also for small-scale helium liquefaction beginning in the early 1990s, by attaching gas circulating tubes, radiation baffles, and condensing stations along the length of the cryo-refrigerator [13, 17, 26, 29–31].

For maximum liquefaction rate, the room temperature gas fed into the liquefier Dewar should be high purity helium, i.e., "clean helium" [8–10]. Indeed, impurities must be eliminated before liquefaction, as they will freeze inside the circulating tubes or on the baffles and condensers attached to the cryo-refrigerator cooling stations, drastically reducing the liquefaction efficiency. Thus, the "clean" gas introduced in the liquefier at room temperature should be first precooled down to 4.2 K vapor, before it reaches the second stage cold finger surface. Optimal gas precooling can only be obtained by very efficient heat exchange methods between helium and the cryo-refrigerator. If the precooling of the room temperature gas, down to 4.2 K vapor, occurs just before the vapor reaches the second stage cooling station, considering that the latent heat of helium at 100 kPa (i.e., at the saturated vapor pressure corresponding to 4.2 K), is 21 J/g, and, using a cryo-refrigerator with a cooling power of 1.5–1.8 W available at the second stage cooling station, liquefaction rates of 0.071–86 g/s (49–59 L/Day) should be possible.

However, in practice, it took almost two decades to reach liquefaction rate values of the order of 0.026 g/s (18 L/Day). This is because efforts were solely focused on designs where cryo-refrigerators operated in vacuum, with the gas circulating inside complex heat exchanger tube designs [26], or cryo-refrigerators operated in gas, with the gas circulating around radiation baffles and condensers [30]. Baffles and condensers were added with the intention of maximizing heat exchange between the helium gas flow and the cryo-refrigerator heat stations.

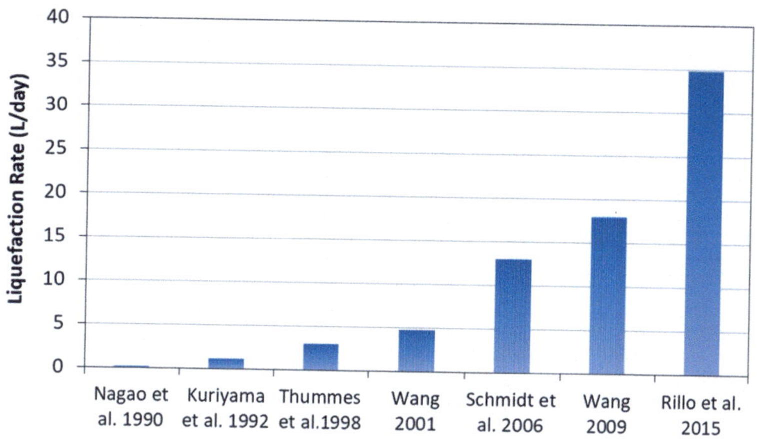

Fig. 10.1 Small-scale helium liquefaction rates reported in literature during the last decades

By applying a new approach to the traditional heat exchange methods described in previous work, we have been able to double the reported maximum liquefaction rates, reaching values exceeding 35 L/Day for a cryo-refrigerator with a nominal cooling power of 1.5 W at 4.2 K [23, 24] (see Fig. 10.1). No heat exchangers, radiation baffles, cooling stations, or condensers are used in this novel approach. Instead a bare cryo-refrigerator inserted in the Dewar neck of a cryogenic vessel, coupled with a precise control of the vapor pressure, to select an optimum pressure (P) and temperature (T) trajectory for the gas in the P–T equilibrium phase diagram during the liquefaction process, is employed. In essence, the method uses elevated pressure to densify and finally condense the vapor at higher temperatures, taking advantage of the larger cooling power available from the cryo-refrigerator stages under these conditions.

10.2 Cryocooler-Based Gas Liquefaction System and Method

10.2.1 Heat Exchange Between Gas and Cryo-Refrigerator

The cooling power that each stage of a closed-cycle refrigerator generates is determined mainly by its temperature, but also depends on the temperature of the other stages. The manufacturer of the cryo-refrigerator generally supplies this information as a two-dimensional load map. For a two-stage cryo-refrigerator (Fig. 10.2), the load map shows the dependence of the power available at the lower second stage cooling station, versus the temperatures of the upper first stage cooling station, as well as the dependence of the power available at the upper first stage cooling station, versus the temperatures of the lower second stage cooling station [5, 27]. Normally, high conductivity copper flanges are provided at the cryo-

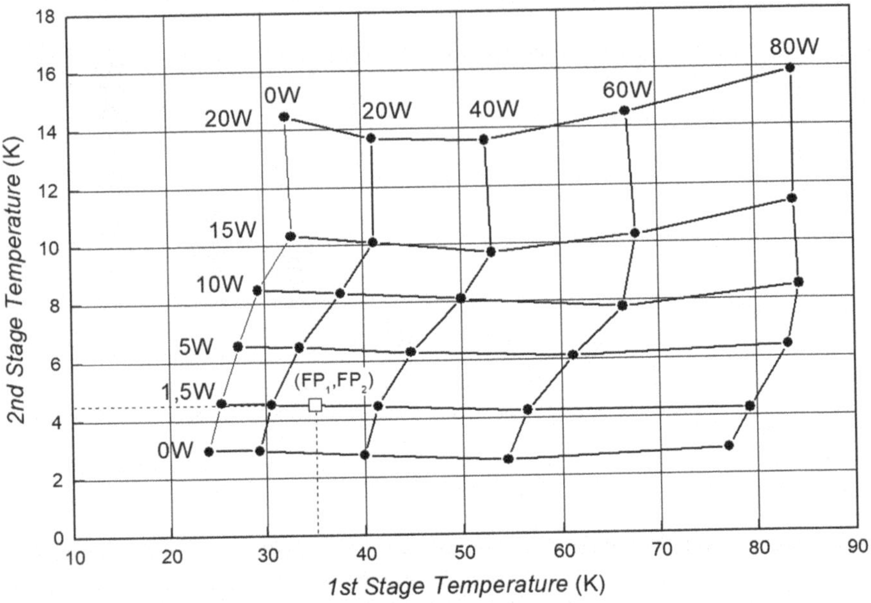

Fig. 10.2 Load map at 50 Hz of a two-stage RDK-415D cryocooler (SHI) in vacuum. Black dots are measured 1st T_1 and 2nd T_2 stage temperatures when different heat power values, FP_1 and FP_2 are applied to the corresponding flanges. The vertical lines join points for specific first stage heat power applied values of 0 W, 20 W, 40 W, 60 W, and 80 W. The horizontal lines join points for specific second stage heat power applied values of 0 W, 1.5 W, 5 W, 10 W, 15 W, and 20 W

refrigerator first and second stages cooling station, as anchor points for thermal contacts with the device to be cooled.

In general, for a cryo-refrigerator with N stages, the cooling power available at each stage generally increases with temperature. The total power that can be extracted from each stage by heat exchange all along its length (regenerator length and expansion volume length) is the integrated power (IP_i) of the stage i, with i ranging from 1 to N. This is higher than the power available at the lower (colder) end flange (Flange Power, FP_i in the following), i.e., $IP_i/FP_i > 1$, due to the capacity of the regenerator to remove heat at intermediate temperatures [22]. In standard applications, in which the cryo-refrigerator operates in vacuum (Fig. 10.3) only FP_i is used. On the other hand, when heat exchange is allowed to occur all along the cryo-refrigerator length with a good thermal conductor such as helium gas (i.e., when IP_i is used for gas precooling before condensation), the liquefaction power, LP, available at the last stage colder end, stage N, will be lower than the FP_N specified on the load map (i.e., $LP/FP_N < 1$) [30, 31]. However, the total power extracted from it, i.e., the power used to precool the gas, IP_N, plus the liquefaction power, LP, is always much higher than FP_N. In fact $IP_N+LP = k\ FP_N$, with k approaching values in the range of 4–8, as we will show in the theoretical calculations.

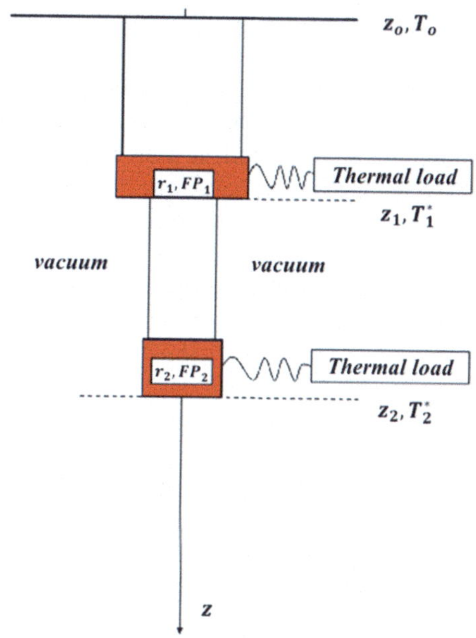

Fig. 10.3 Schematic representation of heat exchange in a cryo-refrigerator standard application, i.e., in vacuum. The power available (FP_1, FP_2) is extracted by solid conduction at each stage flange. The difference between T_1 and T_1* is not significant and it depends on the quality of the thermal anchoring

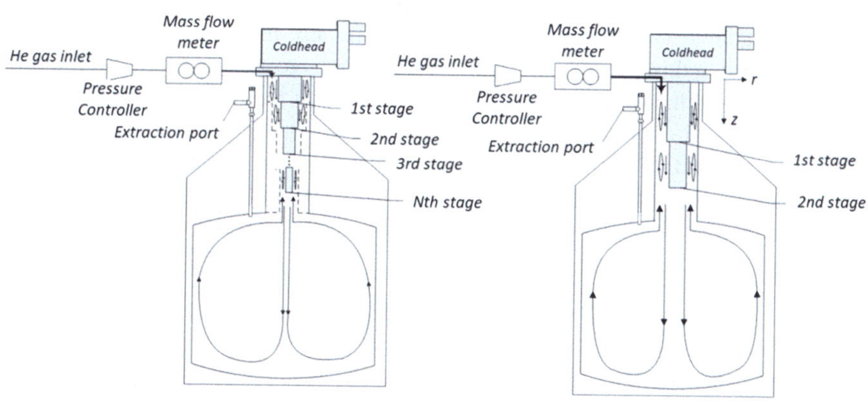

Fig. 10.4 Helium liquefaction system and initial cooling gas convection paths with *N* (left) and with 2 (right) stages refrigeration cryo-refrigerator inside the neck of the Dewar

10.2.2 Liquefaction System and Method

Figure 10.4 shows a schematic drawing of the experimental helium liquefaction system described in this work. It consists of a Dewar with a N stages refrigeration cryo-refrigerator in the neck, a gas inlet module consisting of a pressure controller and a mass flow meter, an extraction port, a liquid level meter (not shown in Fig. 10.4), and finally, N+1 thermometers (i.e., diodes) to measure the temperature of the vapor at each of the cryo-refrigerator expansion volumes (cooling stations) and of the Dewar belly bottom. The initial cooling gas convection paths inside the neck of the Dewar and in the belly are also indicated. The cryo-refrigerator stages are defined by a step pattern, with the cylindrical diameter of each stage being of larger diameter than that of the next stage down in the neck. The neck can be made straight or stepped until it joins with the belly (interior of the isolated container). Since no advantages have been found with stepped necks, in what follows we will describe a system with straight neck. The gas is precooled to its condensation temperature at the lower end of the last stage, by direct thermal contact with the cryo-refrigerator outer surface, thus using IP_i with $i = 1$ to $N - 1$, $IP_N + LP$, and, then the liquid is collected in the Dewar belly.

10.2.2.1 Heat Exchange Process During Initial Cooldown

The liquefier Dewar neck is connected to a pure helium gas reservoir and the pressure in the Dewar is maintained at a constant value by means of a precise electronic pressure controller. When the cryo-refrigerator is turned on, as the cryo-refrigerator stages inside the neck get colder, the gas in the neck gets also colder increasing its density. A positive mass flow from the gas reservoir through the neck to the Dewar belly takes place to compensate for the temperature reduction, since the available volume for the gas and the pressure inside the Dewar are kept constant. During this process thermal gradients are created from the Dewar top flange throughout the neck till the Dewar bottom end of the inner container (Dewar belly). Cylindrical coordinates, z and r, can be used to describe the position of a differential amount of gas inside the neck and belly. The origin of the reference system, $z = 0$, is set at the Dewar top plate flange at room temperature (RT). Since the bottom end of the belly, $z = z_b$, is initially also at RT, the axial temperature gradient in the belly $\frac{dT}{dz}$ will be positive, i.e., the gradient has the direction of the gravitational force, and gas convection will occur, contributing very efficiently to the cool-down process of the belly and of radiation shields in the Dewar inner vacuum chamber. Figure 10.5a, b and c illustrate the first cool-down process by gas convection paths for a liquefaction system in which the neck of the Dewar has a simple straight pattern. The continuous lines with arrows represent the gas convection flow in the interior of the Dewar belly as well as between the neck and the cryo-refrigerator stages.

The initial cooldown from room temperature for a 160 L Dewar with a two-stage Gifford-McMahon refrigerator is illustrated in Fig. 10.6, where the time evolution of temperatures at the end of each stage (T_1 and T_2) and at the Dewar bottom (T_3) are shown.

In the cooling process we can clearly distinguish three regions corresponding to different heat exchange mechanisms. While $T_1 < T_2 < T_3$ $\left(\frac{dT}{dz} > 0\right)$ in region 1, the heat exchange is governed by convection (Fig. 10.5a). In region 2 (Figs. 10.5b and 10.6) where $T_1 > T_2$, helium gas is stratified around the cryo-refrigerator, and

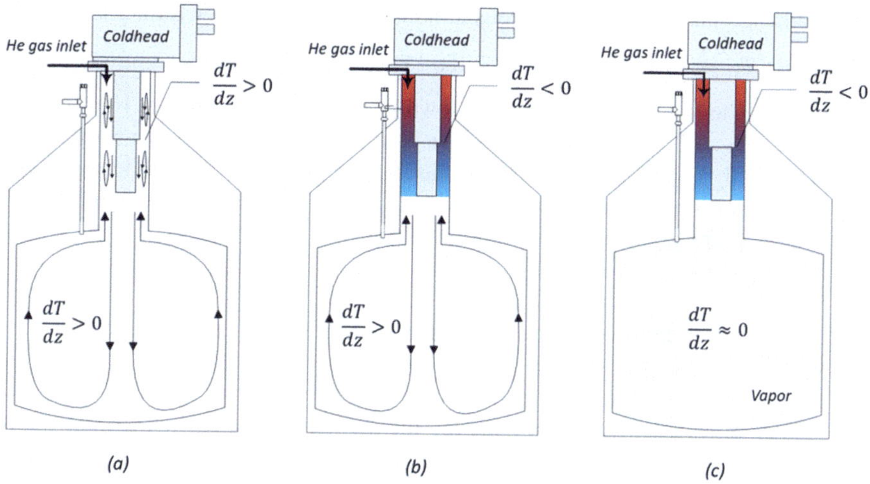

Fig. 10.5 Evolution of the heat exchange mechanisms in the liquefaction system during the first cool-down process from room temperature. The heat exchange takes place by convection (**a**), convection and conduction (**b**), and conduction (**c**)

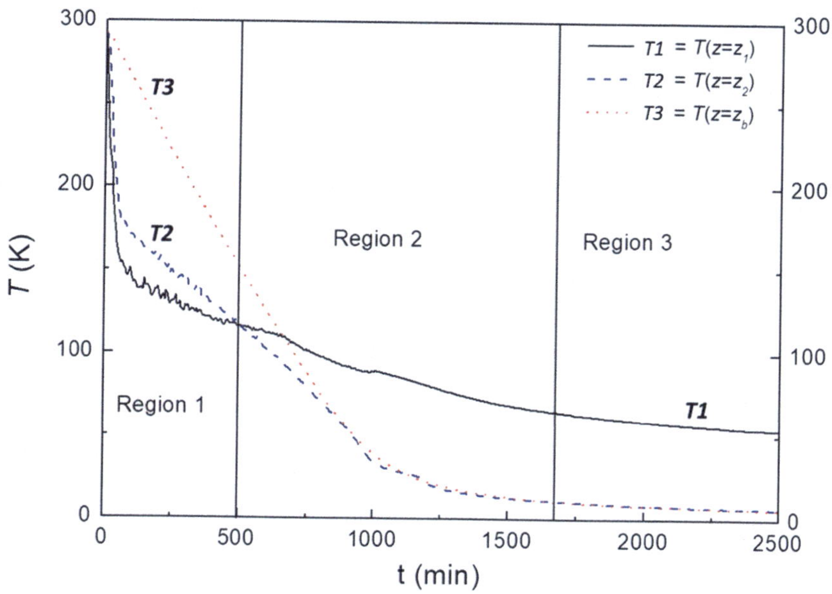

Fig. 10.6 Cool-down curves of a liquefaction system with a two-stage Gifford-McMahon refrigerator and Dewar with straight neck: first stage temperature ($T_1 = T(z = z_1)$), second stage temperature ($T_2 = T(z = z_2)$), and Dewar bottom ($T_3 = T(z = z_b)$). Regions 1–3 correspond to the different heat exchange mechanisms involved in the cool-down process

heat exchange takes place by gaseous conduction through the helium gas. Still in this region $T_2 < T_3 \left(\frac{dT}{dz} > 0\right)$, and, the Dewar belly continues being cooled by convection. The last region of the cool-down process (region 3 in Fig. 10.6) occurs when $T_1 > T_2 > T_3$, where the helium gas is stratified around the cryo-refrigerator and in the belly (Fig. 10.5c), so that the heat exchange occurs by conduction until the liquefaction process starts when $T_2 = T_3 = T_e$, where T_e is the equilibrium temperature corresponding to the equilibrium liquefaction pressure, $P = P_e$.

10.2.2.2 Heat Exchange Process During Liquefaction

In this system, the gas mass flow rate is measured by means of an external mass flow meter attached to the gas inlet of the liquefier Dewar. In the first cool-down process from room temperature, the inlet mass flow oscillates initially with increasing amplitude. Subsequently, the helium flow stops oscillating in favor of a rapid increase through a maximum. This behavior can be modeled using the ideal gas state equation applied to the characteristic time evolution (t) of the temperature for this system in the form $T_3(t) = 4.2 + 293e^{-at}$, which correctly reproduces the curve T_3 shown in Fig. 10.7.

The maximum flow rate is followed by a quasi-steady state flow until the instant at which the temperature of the colder end of the second stage, T_2, and that of the vapor around reaches the equilibrium temperature (i.e., $T_2 = T_e$), corresponding to the equilibrium liquefaction pressure, $P = P_e$. At this temperature an inflexion anomaly occurs in the mass flow rate and a lower and steady flow is attained. This marks the onset of condensation of liquid under the equilibrium conditions (P_e, T_e). The decrease in flow rate at $T_2 = T_e$ is due to the fact that the vapor extracts the latent heat of condensation from the cryo-refrigerator available power, thus resulting in a lower IP_2 for precooling.

At the onset of condensation and under equilibrium conditions (T_e, P_e), a state of constant mass flow ($\frac{dm}{dt}$ = constant) is established. At the same time, precooled vapor is converted into liquid, $\frac{dm_V}{dt}$. Temperature measurements of the gas along the neck, on the cryo-refrigerator stages and on the Dewar belly indicate that, for a given z coordinate the temperature does not depend on r, indicating the absence of measurable radial gradients. On the other hand, the temperature profile along z indicates that helium gas is stratified from the top plate flange, that is at RT, with $\frac{dT}{dz} < 0$ in the neck, along the entire length of the refrigerator, and, till the last stage flange (cooling station) of the cryo-refrigerator. In addition, $\frac{dT}{dz}$ is almost negligible from the second stage to the bottom of the Dewar belly where helium is in liquid form, being this the coldest point inside the inner Dewar container.

Figure 10.8 illustrates the thermal gradients during the liquefaction process under equilibrium vapor–pressure (P_e, T_e) conditions, when the second stage flange has

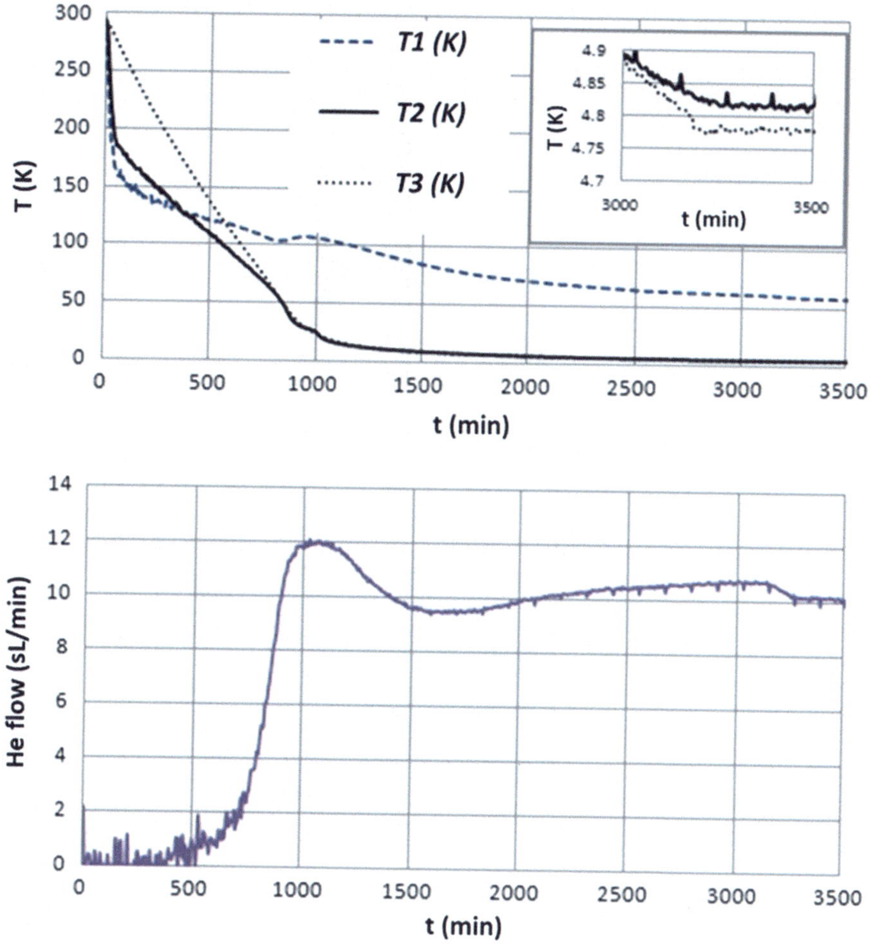

Fig. 10.7 Upper graph: T_1, T_2, and T_3 temperature time evolutions during initial cooldown. The inset shows T_2 and T_3 at the onset of the of liquefaction process. Lower graph: helium gas flow rate during the first cool-down process, from room temperature down to the liquefaction temperature

reached the condensation temperature ($T_2 = T_e$). The temperature gradient along the two-stage length of the cryo-refrigerator has been represented by a colored area, from RT (red) to a condensation temperature T_e at the lower end flange (blue) of the second stage. Since, in this liquefaction method, the gas pressure in the Dewar is set at the saturation pressure of the liquid at a given temperature, as soon as liquefaction starts, the temperature of gaseous and liquid helium in the belly is solely defined by the helium saturation vapor–pressure curve (Fig. 10.9a). The temperature of liquid helium increases with its pressure along the vapor–pressure curve. In the case of ^{4}He, as the pressure increases from around 12 to 227 kPa of absolute pressure, the temperature increases from 2.5 K, to the critical point at 5.19 K. Normally, with

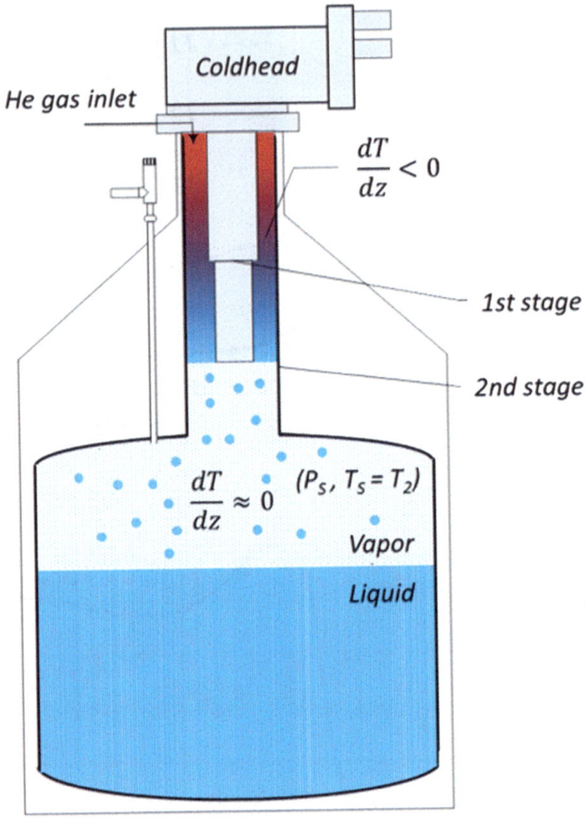

Fig. 10.8 Illustration of the gas temperature stratification during liquefaction process with a two-stage GM refrigerator inside a straight neck of a Dewar in equilibrium conditions (P_e, T_e) during liquefaction process

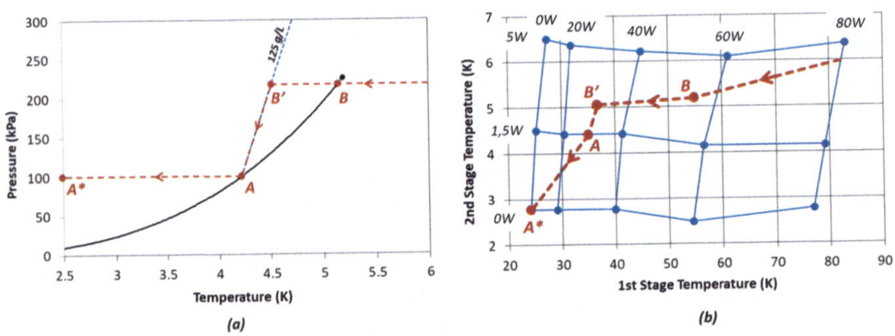

Fig. 10.9 (**a**) Phase diagram of helium-4 and liquefaction trajectory in the range of operation of a typical closed-cycle refrigeration system (2.5–5.192 K, 12.5–227 kPa). (**b**) Experimental liquefaction trajectory superimposed to the RDK-415 two-stage cryo-refrigerator load map of Fig. 10.2. A and B are liquefaction points at two different saturation pressures. B corresponds to liquefaction near the critical point and A corresponds to liquefaction near atmospheric pressure. B' and A* are single-phase overcooled liquid points at the corresponding pressures of the saturation points B and A

no applied load, the lowest temperature reached by closed-cycle refrigerators is about 2.5 K for which the vapor pressure of helium is about 12 kPa. Therefore, a practical range over which the capabilities of closed-cycle refrigeration systems and the helium vapor–pressure curve overlap is from about 12 kPa at 2.5 K to 227 kPa at 5.19 K (see Fig. 10.9a). Obviously, the refrigeration system can also function at intermediate points like A, i.e., at atmospheric pressure (100 kPa) and temperature of 4.23 K. An example of a liquefaction trajectory is illustrated in Fig. 10.9a and b. Starting at high temperature, the gas is cooled at a constant pressure of 210 kPa, down to the saturation temperature at point B, where liquid starts to be produced. The liquefaction system remains at the P_e–T_e conditions (210 kPa, 5.1 K) of point B until the Dewar is full with two-phase saturated liquid helium.

When this occurs the liquid starts to be overcooled (B→B′) and becomes single-phase. Simultaneously, thermal conduction due to the direct contact of the liquid with the cryocooler second stage overcools the coldhead first stage. Thus, the available IP_1 power of the first stage decreases progressively until the system reaches point B′ (see Fig. 10.9b), at which the liquid has the desired density (i.e., the density of saturation condition at atmospheric pressure, that is, the desired pressure for the transfer to experimental cryostats).

From B′ to A there is no mass flow into the liquefier Dewar until the liquid reaches the transfer conditions (100 kPa, 4.2 K). Finally, before the liquid is needed, it can be overcooled down to the base temperature of the cryo-refrigerator, i.e., down to 2.5 K, while keeping the vapor pressure at 100 kPa, that is, being ready for transfer at any moment (point A* in Fig. 10.9(b)).

The transfer of overcooled single-phase liquid can be very efficient, reducing the transfer losses to less than 3% instead of the typical 25% losses of saturated two-phase liquid [23].

Thus, as illustrated by this example, the condensation temperature and hence the power of the last stage of the refrigerator where condensation occurs, can be controlled. Thus, by just controlling the pressure in the Dewar, this simple system allows control of the operating point of the refrigerator, i.e., of the available power IP_i at each stage of the refrigerator and the condensation power LP, as determined by the temperatures of all its stages. Thereby, the heat potential energy that can be extracted from the gas, both for its precooling from room temperature to the point of operation of each cryo-refrigerator stage (IP_i), as well as for its liquefaction (LP), can be perfectly controlled, and is higher the higher is the Dewar pressure. This fact together with the faster enthalpy reduction of helium near and above the critical pressure has allowed a dramatic liquefaction rate enhancement, yielding values in excess of 34 L/Day without need of complex heat exchangers, condensers and radiation baffles [23].

10.2.2.3 Advanced Technology Liquefier (ATL)

Based on the above ideas a commercial version of the helium liquefier was developed (Fig. 10.10). This is the advanced technology liquefier (ATL) with two

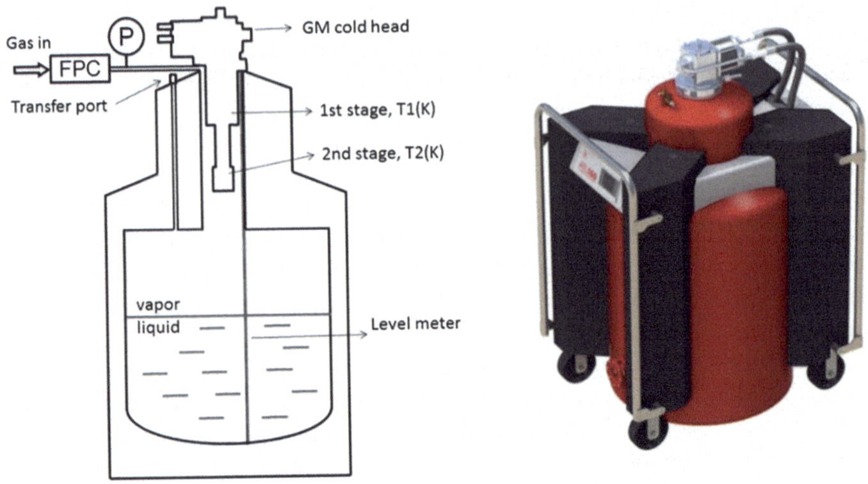

Fig. 10.10 Left: Schematic representation of an ATL160 liquefier. The pressure P of the helium inside the liquefier is controlled by means of a forward pressure controller (FPC). A programmable logic controller (PLC) reads T_1, T_2, and the liquid level meter probe. The PLC sets the desired liquefaction pressure and continuously reads actual P and helium mass flow values from the FPC. Right: ATL160 photography

models of 160 and 80 L capacity [18, 19]. The ATL liquefies and stores gaseous helium supplied from an external source. Helium gas can be provided from ultra-high purity (UHP) helium bottles or recovered from liquid helium boil off. The ATL can liquefy helium gas directly from an instrument in direct recovery mode, or from stored high purity helium gas in liquefy slow, medium, and fast modes [8]. Purchased ultra-high purity bottles or recovered and purified helium gas can provide the necessary high purity gas to feed the ATL. In direct recovery mode the ATL is connected directly to the helium boil off of a system and liquefies the boil off as it enters the ATL at low pressure. When liquefying stored helium gas, the ATL Dewar is pressurized up to 10 psi (in "Fast mode") to maximize liquefaction rates [8].

Figure 10.11 shows experimental results of gas flow, liquefied volume, and temperature of liquid produced as a function of time, in a real ATL160 at three different pressures (107, 135, and 170 kPa). The corresponding liquefaction process points A, B, B′, and A* indicated in Fig. 10.11 are equivalent to those shown in Fig. 10.9.

To transfer helium the ATL must be pressurized below 3 psi. The ATL will reduce the pressure automatically once the Dewar is full. If the user needs to extract liquid sooner, the pressure can be reduced by activating liquefy slow mode and waiting for the pressure to decrease or by opening a manual exhaust check valve.

Once the Dewar is depressurized, a transfer line can be inserted into the ATL and a transfer mode selected. All transfer modes are equivalent, except for the pressure at which the ATL is regulated. Higher pressures transfer liquid helium at a faster rate than lower pressures. The ATL uses a small heater or external high purity gas source

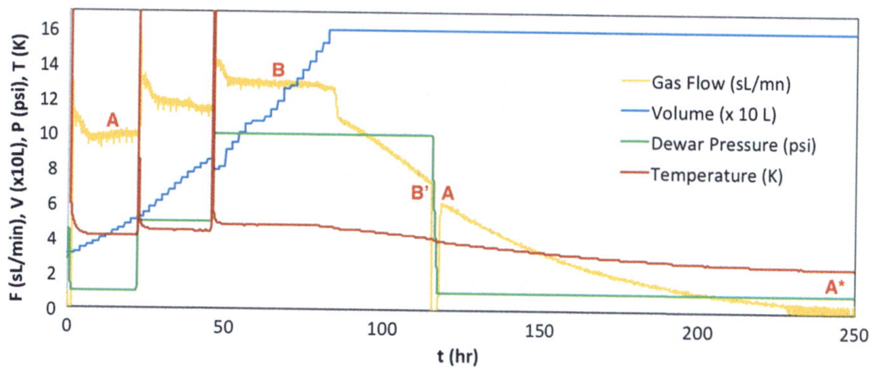

Fig. 10.11 Experimental results obtained in an ATL160 at three different pressures (107, 135, and 170 kPa). Yellow: gas flow rate. Red: Temperature of second stage cold finger and of liquid. Green: Liquefaction pressure. Blue: liquefied volume. A, B, B′, and A* represent liquefaction points like those shown on the *P*–*T* phase diagram and T_1–T_2 load map of Fig. 10.9

to pressurize the Dewar for transfers. To use an external gas source, an ultra-high purity helium bottle or equivalent helium gas source is connected to the high purity helium input on the back of the ATL.

10.2.2.4 Advanced Technology Liquefier (ATL) with Pressurized Insert

In the previous paragraphs, we have described an efficient procedure to extract more cooling power from the cryocooler, following, at elevated pressure around the critical pressure, a path along which the enthalpy decreases much faster with temperature than at around atmospheric pressure. However, increasing the working pressure of the Dewar above a certain limit has two practical disadvantages:

- The liquefier Dewar must be legally certified yearly, if the pressure–volume product exceeds a certain safety criterion, and this will be impractical. With the current regulations this limits the pressure to 170 kPa for a 160 L volume.
- The pressure of the Dewar to transfer liquid helium efficiently should be close to atmospheric pressure, e.g., 107–114 kPa. If a transfer is needed during a liquefaction run at high pressure, it will be necessary to lower the Dewar pressure before performing the helium transfer. This time can be considerable, as it takes around 6 kPa/h (e.g., 10 h to decrease from 170 kPa). It can be performed overnight. In case of need, a high-pressure transfer can be performed but the transfer losses will be very high (of the order of 50%).

To overcome the above problems, while taking advantage of the liquefaction at elevated pressure around the critical pressure, we have developed a new liquefier model [20]. As shown below, the new model fulfills legal regulations, and liquid helium transfers at atmospheric pressure, can be performed at any moment.

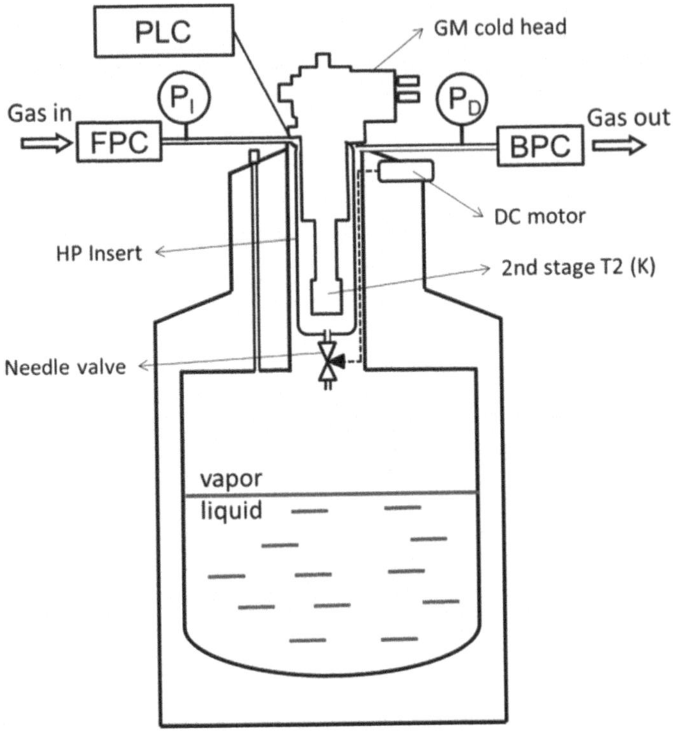

Fig. 10.12 Schematic representation of the ATL160 prototype with pressurized insert. The pressure of the helium inside the insert (P_I) is controlled by means of a forward pressure controller (FPC), and a back-pressure controller (BPC) controls the pressure inside the Dewar (P_D). The helium leaves the insert and expands into the Dewar belly through a needle valve managed by a DC motor. All the electronic components are controlled by a PLC

The idea consists of performing the liquefaction process at a pressure greater than 170 kPa, inside a small neck insert that includes the cryocooler. From this insert, the liquid, or supercritical fluid if the pressure exceeds the critical pressure, is expanded into the Dewar belly through a needle valve. The Dewar pressure is maintained at 107 kPa using a back-pressure controller, which means that helium is released when necessary to the recovery system. Initial experiments were performed on a special prototype based on the standard model ATL160. The schematic of the system is shown in Fig. 10.12.

The main differences in the hardware compared to the ATL described in the previous section are: the liquefaction insert, the needle valve controlled by the DC motor, and the back-pressure controller to maintain the pressure near atmospheric pressure values (107 kPa). All the others components are the same.

In Fig. 10.13 we can see a detailed view of the main elements involved in the liquefaction process. The liquefaction pressure inside the insert is fixed above the critical point at, e.g., $P_I = 230\,\text{kPa}$ by the forward pressure controller (FPC). In

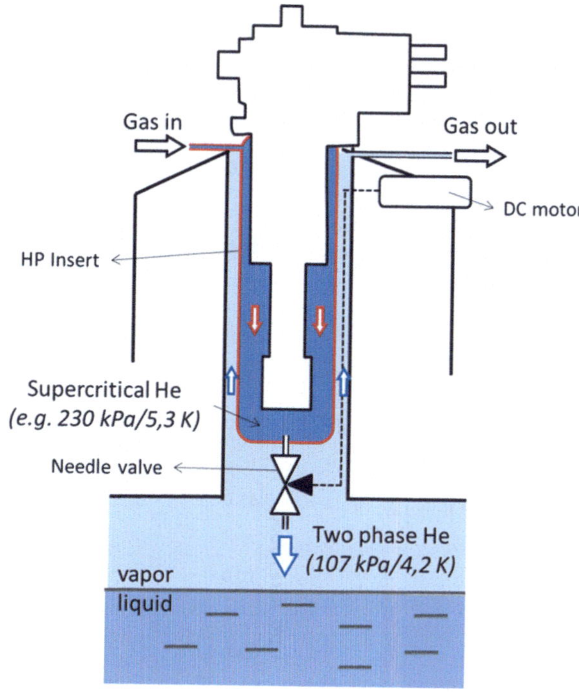

Fig. 10.13 Schematic view of the ATL with pressurized insert in the Dewar neck. The gas at pressure (P_I, e.g., 230 kPa) above critical pressure (dark blue) enters inside the insert (red pipeline). Then it is cooled by the cryocooler. At the end of the cryocooler's second stage the supercritical He is at temperatures around $T_2 = 4.5$–5 K. A needle valve managed with a DC motor allows a constant flow to the Dewar belly ($P_D = 107$ kPa, $T_D = 4.2$ K). Finally, the Dewar belly pressure is maintained near atmospheric pressure (107 kPa) releasing gas through the neck Dewar to the recovery system

this way, as we explained in previous section, we were able to extract more cooling power from the cryo-refrigerator. The pressure in the insert is fixed by the FPC at P_I (above the critical point), the supercritical fluid is then expanded through the needle valve. The control of the valve aperture is made with a proportional-integral-derivative controller (PID controller). The feedback signal for the PID control is the helium input flow through the FPC (F_{IN}), thus, the needle valve aperture will be the necessary to achieve the target flow objective. On the other hand, the helium output flow through the BPC (F_{OUT}) is not fixed. It will be automatically adjusted by the BPC to maintain the Dewar helium bath at the desired low pressure (P_D).

With a good heat exchange for small output flow, the helium that escapes the Dewar will exchange its enthalpy with the input flow, from 107 kPa and 4.29 K to the upper neck region at 107 kPa and 295 K. In other words, the helium gas will be thermalized at ambient temperature (295 K) when it exits the Dewar while precooling the input gas to be liquefied in the HP insert (i.e. it will give back its enthalpy to the input gas, providing extra cooling power).

It must be noted that in the standard ATL, the input flow is determined naturally by the liquefaction pressure [23]. Here, the input flow can be modified with the needle valve aperture, therefore we need to optimize the point (liquefaction pressure and input gas flow), where it is possible to maximize the liquefaction rate while minimizing the helium losses to recovery.

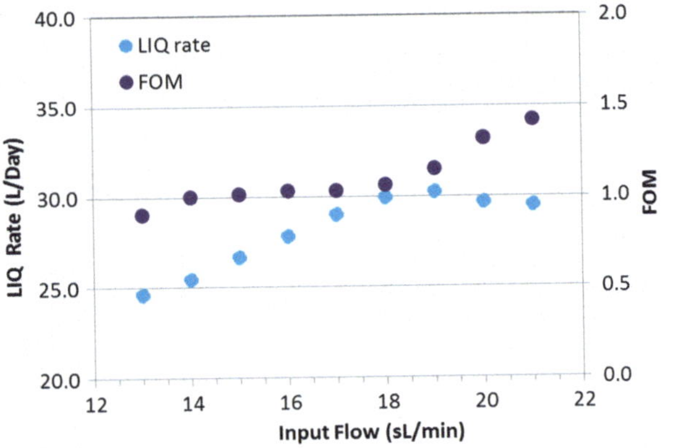

Fig. 10.14 Liquefaction rate (liters of liquid helium at atmospheric pressure produced per day) as a function of input gas flow, for constant input pressure ($P_I = 227.5$ kPa). FOM is a figure of merit defined as the % of output flow with respect to input flow, divided by the liquefaction rate

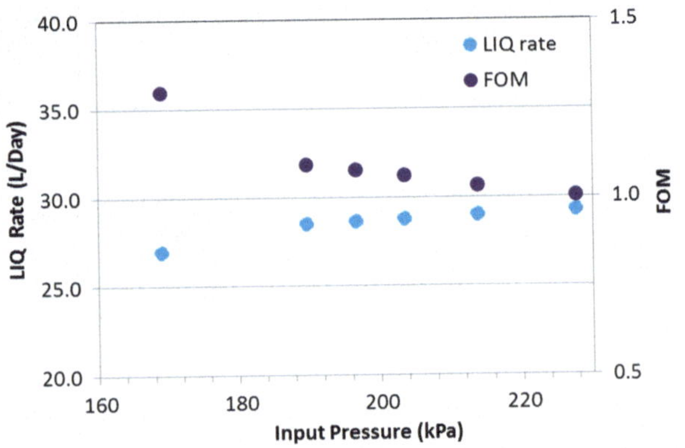

Fig. 10.15 Liquefaction rate and FOM as a function of input pressure (P_I), for constant input flow rate of 17 sL/min

To evaluate the optimal operation point (input flow target) arising from different scenarios, we define the Figure of Merit (FOM) as the % of output flow with respect to the input flow, divided by the liquefaction rate:

$$\text{FOM} = \frac{\% F_{\text{OUT}}}{\text{LiqRate(L/Day)}}. \tag{10.1}$$

Obviously, the optimal situation will be a compromise between liquefaction rate, which should be the maximum possible, and a reasonable FOM (e.g., around 1%/(L/Day)) for the chosen liquefaction rate.

Figure 10.14 clearly shows that there is a region in the range 17–19 L/min of input flow in which the liquefaction rate has a maximum up to above 30 L/day, while the FOM does not exceed much from 1%/(L/day), a value considered reasonable.

To complete the study, it will be interesting to know how the liquefaction rate and FOM depend on the input pressure (P_I) at a fixed input flow rate.

We selected a fixed input flow rate of 17 sL/min, and then we varied the input pressure. Figure 10.15 shows that the liquefaction rate decreases with decreasing input pressure, and FOM increases. The higher increase in the liquefaction rate is produced at lower pressures and once we are close to the critical point (227.61 kPa), it seems to tend asymptotically to 30 L/day. Therefore, the best results are obtained when the input pressure is near the critical point.

In summary, we have seen that the input flow and the input pressure determine the liquefaction rate and FOM. The FOM can be maintained in a reasonable value of the order of 1%/(L/Day) or even smaller. The best results (LIQ rate = 30 L/day, and FOM = 1%/(L/Day)) are obtained for an input pressure near the critical point (P_I = 227.6 kPa), and an input gas flow of 18 sL/min [8]. The commercial version of the ATL with pressurized insert is the ATL160+ [20]. It uses an RDK-415 in combination with a smart energy compressor [3, 7] to optimize the efficiency of this small-scale liquefaction new technology that approaches that of Collins large scale liquefaction systems [4].

10.3 Cryocooler-Based Gas Purification System and Method

As already mentioned, a high efficiency in helium liquefaction requires the gas to be very pure. In this section we describe a purification technique also based on cryocoolers.

10.3.1 Gas Purification by Cryo-condensation

Let us introduce first, a general phase diagram for a pure substance. Three phases are possible: solid (S), liquid (L), and gas (G). The phase diagram delimitates the regions of each phase in the pressure/temperature diagram. The coexistence lines of the two phases: S-L, L-G, and S-G, the triple point (S-L-G coexistence), and the critical point, are also indicated. Furthermore, the critical point marks the zone from which the substance is a supercritical fluid.

Figure 10.16 shows the different paths, at constant pressure, between phases, e.g., sublimation (S⟶G), condensation (G⟶L), deposition (G⟶S). Let us consider the gas to be purified and its impurities as a mixture of gases. If the concentration of impurities is small enough, by reducing the temperature of the mixture at constant pressure, the partial pressures of impurities (and, consequently, their concentration in the mixture) will remain constant until they reach the deposition line. Afterwards, by further cooling of the gas mixture, there will be a dramatic reduction of the impurities partial pressures (and of their concentration in the gas), which will become negligible at low temperatures. In a practical case, we consider that the initial molar fraction, y_j, at room temperature (RT), of an impurity represented by the index "j" in the gas mixture, can be approximated by the ratio of its partial pressure, P_j, to the total pressure of the mixture, P_t [2]. This approach is strictly valid only for ideal gases and small molar fractions. Since the recovered helium is

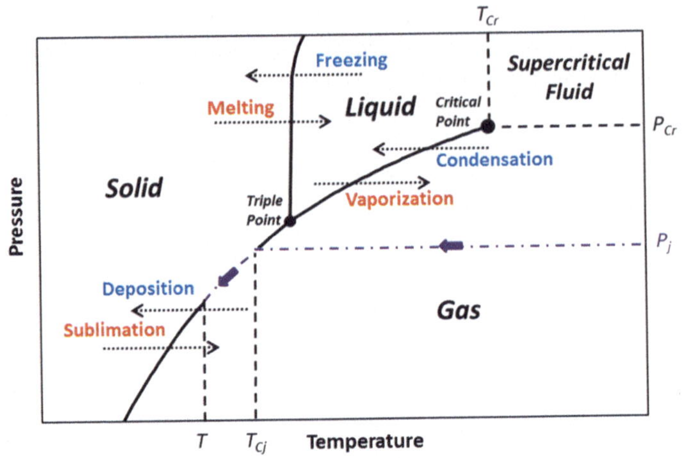

Fig. 10.16 Model phase diagram for a pure substance with the different phases separated by solid lines. The horizontal dotted arrows indicate phase transitions at constant pressure. The purple line represents a trajectory of a certain quantity of this substance cooled below its condensation temperature (T_{cj})

composed of helium gas with small quantities of impurities (<2%), it is reasonable to consider that the following expression can be applied:

$$y_j = \frac{n_j}{n_t} = \frac{P_j V/\mathrm{RT}}{P_t V/\mathrm{RT}} = \frac{P_j}{P_t}. \tag{10.2}$$

The partial pressure of a frozen impurity at any temperature, T, below its condensation temperature, T_{cj}, that is, for any $T < T_{cj}(P_j)$, is given by the vapor pressure of the condensate at T; in other words, it can be represented by the solid line separating gas (G) and solid (S) phases for the specific impurity.

Now we will consider the practical case of small amount of impurities (<2%) in helium. The simple inspection of Fig. 10.16 shows that $y_j(T)$ for any possible impurity dramatically decreases by orders of magnitude once the sublimation (G$\longrightarrow$S) line is reached and T is further decreased.

As it is well known [1], the impurities in recovered helium come mostly from the atmosphere; therefore, they are basically nitrogen and oxygen, with a small number of other constituents (H_2O, CO_2, CO, etc.). When a mixture of helium gas and its typical impurities, with total constant pressure P_t, is cooled down for purification using cryo-condensation, the partial pressure of any impurity in the vapor phase, $P_j(T)$, remains constant with T until the condensation saturation-line of the impurity is reached. This occurs at a temperature T_{cj} that depends on the initial molar fraction y_j (300 K). In other words, above T_{cj} the $P_j(T)$ values can be approximated by the product of the initial molar fraction in the vapor phase, and the total gas pressure [2]:

$$P_j(T) = y_j(300\,\mathrm{K}) \cdot P_t, \quad T > T_{cj}. \tag{10.3}$$

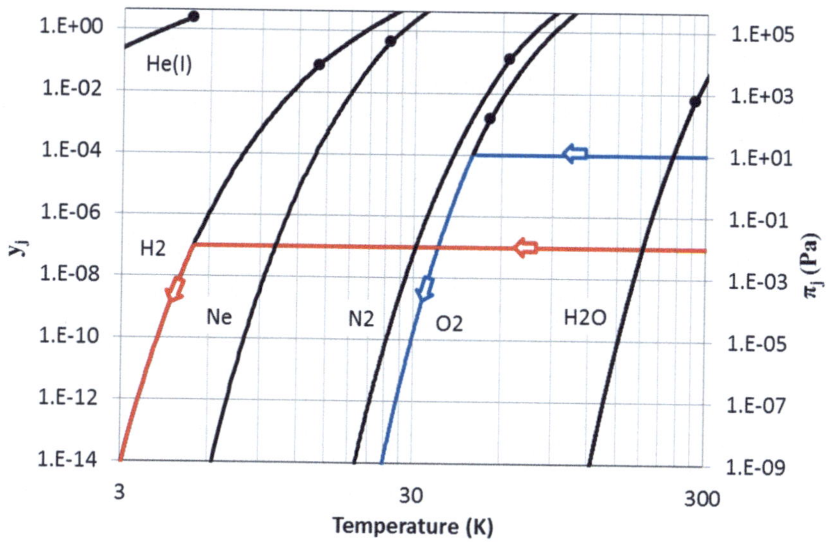

Fig. 10.17 Molar fractions $y_j(T)$ and partial pressures $n_j(T)$ of O_2 (blue) and H_2 (red) in recovered helium as a function of T, for initial molar fractions of 10^{-4} and 10^{-7}, respectively, at STP (100 kPa and 273.15 K). Black solid lines are the equilibrium vapor pressure P–T line of He(I) and the vapor–pressure lines of some typical impurities $\pi_j(T)$ as a function of T, in the range 3–300 K. When the impurity reaches the G→S line, it follows the continuous line, down in T, and does not leave this line

Once the corresponding vapor–pressure saturation-line of the impurity is reached at T_{cj}, the partial pressure of the impurity in the vapor phase coincides with its vapor pressure $\pi_j(T)$,

$$P_j(T) = \pi_j(T), \quad T \leq T_{cj}. \tag{10.4}$$

Therefore, once the condensation of a given impurity starts, its molar fraction in the vapor phase is not constant anymore but decreases dramatically with T, because $\pi_j(T)$ also does, and is given by

$$y_j(T) = \frac{\pi_j(T)}{P_t}, \quad T < T_{cj}. \tag{10.5}$$

Figure 10.17 shows the molar fraction variation $y_j\ (T)$ of O_2 and H_2 calculated from Eqs. (10.3) and (10.5), having initial molar concentrations of 10^{-4} and 10^{-7}, respectively, for a helium gas purification pressure $P_t = 100$ kPa. For clarity sake the figure also includes the P–T liquid–vapor saturation-line for the condensed phase of the carrier gas, He (I), and the P–T liquid–vapor and solid–vapor saturation lines for the most frequent impurities typically found in recovered helium, from 3 K up to 300 K [14]. The critical point of He (I) and the triple points of the impurities are marked.

Thus, from a simple inspection of Fig. 10.17 we can conclude that, below 20 K, only Ne and H_2 may have a non-negligible gas molar fraction in He. Furthermore, below 6 K only H_2 gas can flow together with liquid helium, after mechanical filtering. In fact, the presence of O_2 and N_2 molecules in helium can be neglected below 24 K and 20 K, respectively. However, the presence of H_2 molecules cannot be neglected until temperatures well below 3 K are reached [1].

An impurity concentration below the ppm is adequate to guarantee a long-term operation, without significant efficiency reduction, of any liquefier. This purity level can be reached by a recovered helium purifier based in cryo-condensation with a gas circulation that reaches temperatures below 30 K at some point. At the lower temperature region, a mechanical filter guarantees the absence of dragging of any solid impurity. Therefore, the output helium should have a purity level better than 99.9999% (less than 1 ppm in total) for all the contaminants, except Ne and H_2. But, as we will introduce later, there is an impurity that should be reduced below the ppt level to avoid an important issue, the blockage of small flow impedances used in some cryogenic instruments to reach temperatures below 4.2 K. It is the hydrogen.

The hydrogen would be eliminated (<ppt) by cryo-condensation if the temperature were reduced one order of magnitude, from 30 to 3 K.

Thus, we will consider two purifier regimes. One of them, below 30 K, for the atmospheric impurities removal (standard purifier) explained in this chapter, and other, below 3 K or combined with other strategies, for completely hydrogen removal ("Clean Helium" production) [8, 10].

10.3.2 Purifier Design

This section deals with the selection of the minimum elements required to construct a cryocooler-based purified using the cryo-condensation as the principle of operation.

Recalling Eq. (10.5), the value of P_j is given by the amount of the impurity "j" in the gas mixture, being P_t the total pressure of the mixture. The working pressure has to be fixed at a value above atmospheric pressure. The higher the pressure, the lower the impurities content at the output. The pressure is fixed using a forward pressure controller (FPC).

The main impurities in recovered helium are air components (N_2, O_2, CO_2, H_2O); therefore, if we select the working purifier temperature below 30 K, the impurities content will be lower than 1 ppm for all these contaminants (see Fig. 10.17).

In Fig. 10.18 we present the main design elements of the purifier [21].

As we have seen, the pressure inside the Dewar is fixed by the FPC. When the cryo-refrigerator is turned on, the gas mixture will cool down. As the gas is cooled its density increases, so that more gas will enter the system, through the FPC, to maintain the pressure constant.

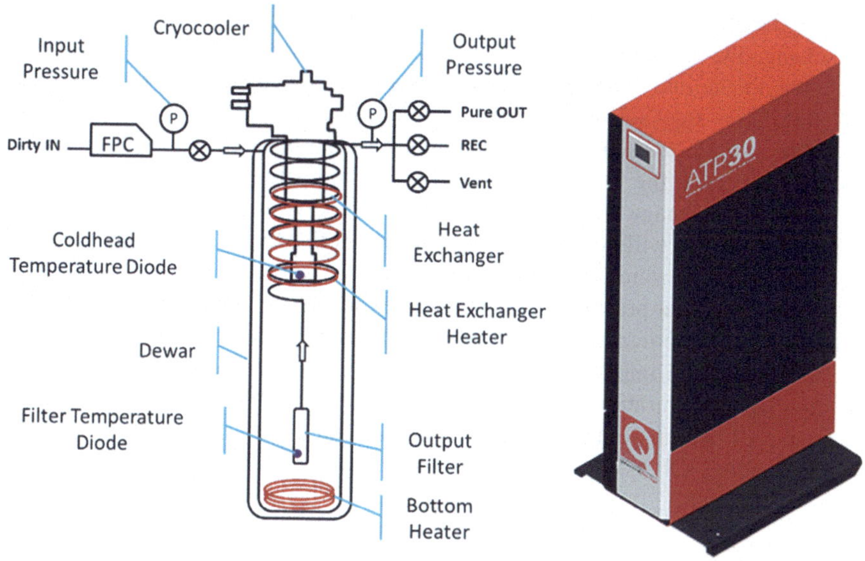

Fig. 10.18 Left: Main elements of the purifier. The "dirty" gas enters into the Dewar through a FPC that fixes the system pressure. The gas is cooled from top to bottom, and it passes through a filter to avoid any drag of solid frozen particles. The output gas exchanges enthalpy in its output path though the heat exchanger, and the clean gas is directed with valves to the proper output depending on the necessity. When the purifier has enough impurities inside, they could be removed quickly using the heaters, and sending the gas to the recovery system or to the atmosphere (depending on the impurities content in each step of the regeneration process). Right photo on an ATP30

The recovered helium, that contains air impurities, enters into the Dewar neck. In this upper region, the temperature is close to RT (293 K). The helium gas descends, and it is cooled, in direct contact with the cryo-refrigerator and the output heat exchanger.

The helium path starts at around 293 K. When the gas reaches the condensation temperature for the component "j" (T_{cj}) (see Fig. 10.17), at some point near the cryo-refrigerator first stage, the component "j" will be solidified by deposition, and will attach to the metallic cold surfaces (cryo-refrigerator or the heat exchanger). After that point, and at lower positions, the gas temperature decreases and the molar fraction in the vapor phase of the component "j" will be lower and lower with T, as $\pi_j\,(T)$ (Eq. (10.5)).

When a region of temperature below 30 K is reached, the helium can be considered already pure enough and it can be directed towards the outlet. To do that, we can use a mechanical filter in the micron range, which will avoid the possible dragging of solid particle impurities to the output.

After the filter, in order to be energy efficient, the clean and cold helium is forced to exchange the enthalpy from 30 to 300 K with the warm and dirty input helium. To do that the helium output path consists of a heat exchanger in the form of a thin walled stainless-steel solenoid tube around the cryo-refrigerator.

Thanks to the heat exchange, the incoming gas is partially cooled by the outgoing gas, and therefore, the required power of the cryo-refrigerator is minimized. So, the system can manage high flows. In addition, the cryo-refrigerator excess power during purification will counteract the growing inefficiencies caused by the solid impurities coating around the cold surfaces.

Let us consider a practical example to have an idea of the order of magnitude of the purifier specifications: the cryo-refrigerator region has a free volume for the storing of solid impurities of the order of 1 L, and the recovered helium has an impurity concentration (y_j) in the order of 0.1% (1000 ppm in total). Then, the recovered helium volume, which the purifier can manage before the impurities storage region is full, is about 1 million of sL:

$$V_{\mathrm{HeREC}} = \frac{1}{y_j} V_{\mathrm{IMP}} R_{G/S}(j) = \frac{1}{10^{-3}} \cdot 1\,\mathrm{L} \cdot 1000 = 10^6\,\mathrm{sL}. \tag{10.6}$$

The volume ratio $R_{G/S}(j)$ used in (10.6) indicates only the order of magnitude. From [14], we can calculate the real ratio for N_2 case at a fixed pressure (e.g., 100 kPa (1 barg)):

$$R_{G/S}(\mathrm{N_2}) = \frac{\rho_{\mathrm{N_2}}(1\,\mathrm{barg}, 63\,\mathrm{K})}{\rho_{\mathrm{N_2}}(1\,\mathrm{barg}, 290\,\mathrm{K})} = \frac{867.30\,\mathrm{g/L}}{1.16\,\mathrm{g/L}} = 746.4 \tag{10.7}$$

If the purifier manages an average gas flow of 20 sL/mn, then the time to fill the storage volume with impurities is around 1 month (34.7 days). The output pressure is monitored with a pressure sensor. Thus, the pressure difference between input/output allows to detect when the storage volume is full (i.e., the solid impurities produce a blockage in the Dewar). When this occurs the purifier needs to be regenerated. Once the system is regenerated, all the impurities collected inside the Dewar have been evacuated, therefore the vessel is clean and the purification cycle can start again.

10.4 Theoretical Calculations of the ATL Liquefaction Process

This section presents analytical models to calculate mass and heat exchange balance during liquefaction in an ATL, with the aim to infer possible improvement strategies.

10.4.1 Mass Balance

We verified in Sect. 10.2 that during liquefaction there is no thermal gradient against the direction of the gravitational force. Then, we can conclude that the liquefaction process occurs without convection exchange. Instead a pure gas thermal conduction mechanism between stratified helium layers is responsible for the heat exchange between the cryo-refrigerator and the gas.

Therefore, warm gas enters the neck at a rate dm/dt and it is immediately stratified in temperature and distributed along the neck to compensate for the volume contraction of the amount of vapor $\frac{dm_V}{dt}$ that is transformed into liquid $\frac{dm_L}{dt}$. Then the mass balance equation is

$$\frac{dm}{dt} = \frac{dm_V}{dt} + \frac{dm_L}{dt}. \tag{10.8}$$

The liquid and vapor volumes, V_L and V_G, respectively, are related through the following expression:

$$V_T = V_L + V_V, \tag{10.9}$$

where V_T is the total volume of the Dewar inner container. Since V_T is constant, the time derivative of Eq. (10.9) gives the relation between the vapor volume change and the liquid mass change during the liquefaction process:

$$\frac{\frac{dm_L}{dt}}{\rho_L(P,T)} = \frac{\frac{dm_V}{dt}}{\rho_G(P,T)}. \tag{10.10}$$

From (10.8) and (10.10) the relation between the liquefaction mass rate and the warm gas input flow rate can be obtained:

$$\frac{dm_L}{dt} = \frac{dm/dt}{1 - \frac{\rho_V(P_e, T_e)}{\rho_L(P_e, T_e)}} \tag{10.11}$$

where $\rho_V(P_e, T_e)$ and $\rho_L(P_e, T_e)$ are the density of the gas and liquid, respectively, at equilibrium conditions of pressure, P_e, and temperature, T_e. The liquefaction rate can be expressed in practical units, e.g., in g/s, or in L/Day dividing the g/s value by the liquid density at equilibrium conditions (P_e, T_e).

Equation (10.11) indicates that at low pressures (i.e., when P<1 bar) for which $\rho_V << \rho_L$, the mass liquefaction rate can be approximated by the input gas mass flow rate. Instead, as the liquefaction pressure increases and approaches the critical value, the density ratio approaches unity ($\rho_L = \rho_V = 70\text{g/L}$) and the mass liquefaction rate diverges (see Fig. 10.19a). This means that, near the critical point,

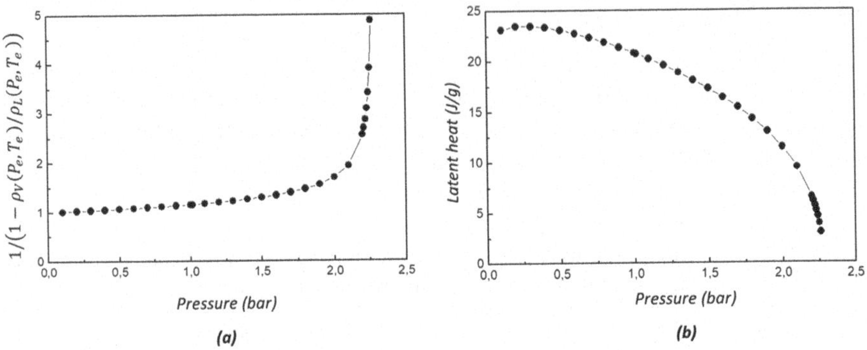

Fig. 10.19 (**a**) The factor $1 - \frac{\rho_V(P_e,T_e)}{\rho_L(P_e,T_e)}$ on Eq. (10.11) by means of which the liquefaction rate, $\frac{dm_L}{dt}$, in equilibrium conditions (P_e, T_e) known the input mass flow, dm/dt, can be calculated. (**b**) Helium latent heat as function of the equilibrium pressure

a small input gas flow rate produces an instantaneous huge liquefaction mass rate, even with a finite small LP available at the condensation flange of the second stage, since the latent heat of He near T_c tends to zero (see Fig. 10.19b).

Even though the instantaneous liquefaction rate of precooled gas could be as high as desired by approaching the critical point (P_c, T_c), the time necessary to precool the gas to the liquefaction temperature dramatically reduces the effective liquefaction rate. On the other hand, most applications require in practice to use the liquid at 1 bar, 4.23 K, for efficient liquid transfer to user equipment. That is, the system has to go from liquefaction point B (near the critical point) to point A on the *P*–*T* phase diagram following the corresponding load map $T_1 - T_2$ trajectory (see Sect. 10.2.2.2, Fig. 10.9a and b). Then, the final gain is represented by the shaded area on the experimental flow rate diagram as illustrated in Fig. 10.20, this is equivalent to 2–3 days less for completion of the liquefaction run from 0 to 100% of the liquid volume capacity at 1 bar [23, 24].

10.4.2 Heat Exchange Balance

In this section, the heat exchange balance during a liquefaction process in a Dewar with a two-stage Gifford-McMahon cryocooler in the neck, is analyzed with the help of Fig. 10.21. On the figure the power distributions, mass flows, temperatures, and enthalpies during the liquefaction process are indicated. The heat power per unit length absorbed by the cryo-refrigerator is defined by $\delta\dot{q}(z)$. T_0 and h_0 are the temperature and the enthalpy of the inlet helium room temperature (RT) gas at $z = 0$. T_1 and h_1 are the temperature and the enthalpy of the helium gas after precooling along the first stage, that is at the first stage colder end, $z = z_1$. The integral power, IP_1, transferred from the inlet helium gas (RT) to along the first

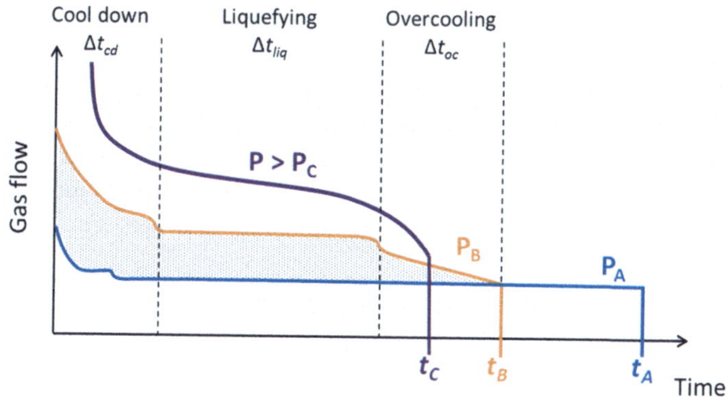

Fig. 10.20 Representation of the typical flow for liquefaction runs at P_A, P_B, and $P > P_C$ the critical pressure of helium in phase diagram of Fig. 10.9a. The average liquefaction rate increase obtained when liquefying at pressure P_B when compared to a liquefaction process at pressure near atmospheric ($P_A \sim 1.1$ bar) is indicated by the shaded region in the diagram

Fig. 10.21 Liquefaction process in a Dewar with a two-stage Gifford-McMahon cryocooler in the neck. In the figure the power distributions, mass flows, temperatures, and enthalpies during the liquefaction process for a two-stage refrigerator are indicated

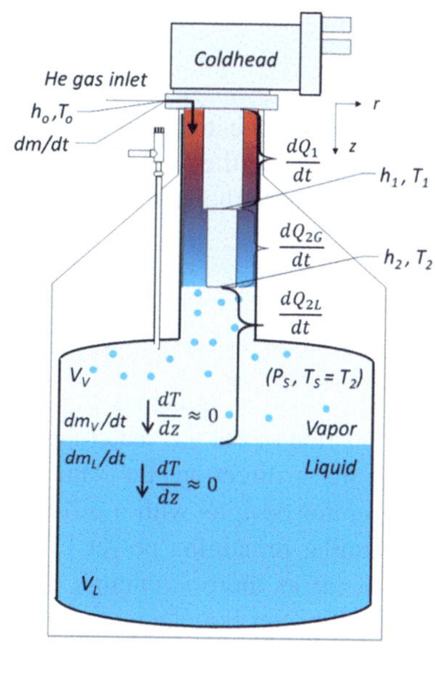

stage length, can be expressed by $IP_1 = \frac{dQ_1}{dt} = \int_{z=0}^{z=z_1} \delta q(z)dz$. T_2 and h_2 are the temperature and the enthalpy of the helium gas after precooling along the second stage length, at the second stage colder end (flange), being T_2 the equilibrium vapor–liquid temperature. Then $IP_2 = \frac{dQ_2}{dt} = \int_{z=0}^{z=z_2} \delta q(z)dz$ is the precooling power transferred to helium gas along the second stage length. LV is the vaporization latent heat of the helium at certain equilibrium pressure and temperature conditions (P_e, T_e). Then, the liquefaction power $LP = \frac{dQ_{2L}}{dt} = \int_{Z=z_2}^{Z=Z_L} \delta \dot{q}(z)dz$ is the condensation energy rate transferred between the precooled helium gas and the colder end of second stage, for the condensation of overheated or equilibrium gas. While T_2

remains above T_c, LP = 0, and all the available power, [IP_1+IP_2] ($T_2 > T_c$), is used for gas precooling (Fig. 10.5, Sect. 10.2.2.1). When T_2 is smaller than Tc the precooling power is reduced by the overheated gas condensation latent heat LP ($T_e < T = T_2 < T_c$) (a first inflexion point on the mass flow rate may be observed). When $T_2 = T_e$, the precooling power is reduced by the equilibrium condensation latent heat LP ($T = T_2 = T_e$), a more important second inflexion point on the mass flow rate is always observed. The energy equations for N stages are

$$\frac{dQ_i}{dt} = \text{IP}_i = \frac{dm}{dt}\,(h_{i-1} - h_i)\,,\, i = 1, N-1 \qquad (10.12)$$

$$\frac{dQ_N}{dt} = \text{IP}_N + \text{LP} = \frac{dQ_{\text{NG}}}{dt} + \frac{dQ_{\text{NL}}}{dt} \qquad (10.13)$$

$$\frac{dQ_{\text{NG}}}{dt} = \text{IP}_N = \frac{dm}{dt}\,(h_{N-1} - h_N) \qquad (10.14)$$

$$\frac{dQ_{\text{NL}}}{dt} = \text{LP} = \frac{dm_L}{dt} L_V \qquad (10.15)$$

Therefore, the total power, $(dQ_T)/dt$, transferred from the gas to the cryo-refrigerator, i.e., the total power extracted from the cryo-refrigerator to precool and then condensate the vapor, will be given by

$$\frac{dQ_T}{dt} = \sum_{i=1}^{i=N} \frac{dQ_i}{dt} = \frac{dm}{dt}\,(h_0 - h_N) + \text{LP}. \qquad (10.16)$$

This equation indicates that, in principle, due essentially to the temperature stratification of the gas matching perfectly the cryo-refrigerator temperature profile along the z direction, the same liquefaction rate could be obtained using cryo-refrigerator designs with a different number of stages. This is true as far as there is a similar remaining power LP for liquefaction at the cold end of the last stage, and as far as there is enough excess power to precool the gas from T_0 (RT) till the equilibrium liquefaction temperature T_e at the lower end of last stage. In other words, a single stage cryo-refrigerator with LP available at its cold end could produce the same liquefaction rate than a two-stage cryo-refrigerator, or than a three stage cryo-refrigerator, or than a four, with the same LP at the second or third or fourth stage lower end, respectively.

Of course, the final cryo-refrigerator design and the minimum required number of stages will be determined by the state of the art of cryo-refrigerator engineering with the required performances.

In the case of a two-stage cryo-refrigerator, Eqs. (10.13)–(10.16) become:

$$\frac{dQ_1}{dt} = \text{IP}_1 = \frac{dm}{dt}\,(h_0 - h_1) \qquad (10.17)$$

$$\frac{dQ_2}{dt} = \text{IP}_2 + \text{LP} = \frac{dQ_{2G}}{dt} + \frac{dQ_{2L}}{dt} \tag{10.18}$$

$$\frac{dQ_{2G}}{dt} = \text{IP}_2 = \frac{dm}{dt}(h_1 - h_2) \tag{10.19}$$

$$\frac{dQ_{2L}}{dt} = \text{LP} = \frac{dm_L}{dt} L_V \tag{10.20}$$

$$\frac{dQ_T}{dt} = \frac{dm}{dt}(h_0 - h_2) + \text{LP}. \tag{10.21}$$

Using the data obtained during different experimental liquefaction processes (the input mass flow rate and the equilibrium pressure), and introducing those values as well as the densities of the gas and liquid on Eq. (10.12), the instantaneous mass and volume liquefaction rates and the average liquefaction rates at 4.23 K can be calculated. From the previous experimental data and the enthalpy dependence on temperature $h(T)$, the integral heat power along each stage and the liquefaction power can be calculated with Eqs. (10.18)–(10.21). Table 10.1 lists the inlet mass flow, the instantaneous and average liquefaction rates, the heat load, the temperature, and the enthalpy, during two different liquefaction processes, for two different

Table 10.1 Inlet mass flow rate, liquefaction rates, heat loads, temperatures, and enthalpies, for two different pressures during liquefaction in an ATL system with a two-stage RDK-415E cryocooler. The more relevant parameters are in bold

P (kPa)	**100**	**210**
dm/dt (g/s)	0.027	0.041
$\frac{dm_L}{dt}$ inst. (g/s) Eq. (10.11)	0.031	0.079
$\frac{dV_L}{dt}$ inst. (L/Day at T_2)	22	73
$\frac{dV_L}{dt}$ avg. (L/Day at 4.23 K)	**22**	**34**
h_0(J/g)	1563.8	1563.8
T_0(K)	300	300
h_l(J/g)	186.6	291.13
$\boldsymbol{T_1}$**(K)**	**35**	**55**
$\text{IP}_1\ (T_1) = dQ_1/dt$(W) Eq. (10.17)	36.9	52.18
$\text{IP}_1\ (T_1)\ /\text{FP}_1\ (T_1)$	36.9/32=1.15	52.18/55 0.95
H_2(J/g)	20.75	16.82
$\boldsymbol{T_2}$**(K)**	**4.23**	**5.1**
$\text{IP}_2 = dQ_{2G}/dt$(W) Eq. (10.19)	4.46	11.3
L_V(J/g)	20.75	9.48
$\text{LP}= dQ_{2L}/dt$(W) Eq. (10.20)	0.64	0.75
$\boldsymbol{\dot{m}/\dot{m}_0 = (\text{IP}_2 + \text{LP})\ /\text{FP}_2}$	(4.46 + 0.64)/1.5 = 3.4	(11.3 + 0.75)/1.5 = **8.0**
$\boldsymbol{dQ_T/dt}$**(W)** Eq. (10.21)	**42**	**64.4**

operating conditions: (A) atmospheric pressure, i.e., 100 kPa, and (B) 210 kPa, and using a two-stage RDK-415E cryocooler from Sumitomo Heavy Industries [27]. These two experimental liquefaction processes correspond to liquefaction at points A (100 kPa, 4.23 K) and B (210 kPa, 5.10 K) of the P–T helium liquid–vapor saturation line in Fig. 10.9, Sect. 10.2.2.2, respectively.

The parameter $\dot{m}/\dot{m}_0$ is the ratio of the actual mass flow rate being cooled and liquefied to the mass flow rate that can be cooled and liquefied using only the 1.5 W cold station heat power at 4.2 K, i.e., the ratio of the total heat power extracted by the helium mass flow rate from the second stage refrigerator, to the 1.5 W that will be available at the cold station under vacuum.

The higher the pressure, the higher the IP extracted along each cryo-refrigerator stage length, and, the higher the power available for condensation, $\frac{dQ_{2L}}{dt} = \mathrm{LP}$. Moreover, according to Eq. (10.21), since the latent heat of helium-4 decreases when the pressure increases, and since the density decreases, the effect of pressure on the instantaneous liquefaction (volume) rate, $[dV_L/dt]_{\mathrm{inst}}$, is very noticeable. Finally, since the total power extracted at 210 kPa and 5.1 K (64 W) is about 1.5 times that of extracted at 4.23 K (42 W), the average liquefaction rate, $[dV_L/dt]_{\mathrm{avg}}$, at 210 kPa is also 1.5 times higher (34 L/Day) than that that obtained at 100 kPa (22 L/Day).

Figures 10.22 and 10.23 present an analysis of the data of Table 10.1 in a graphic way. The blue circles are the first stage heat power load line $P_1(T)$ of a cryo-refrigerator RDK-415 operating at 50 Hz, in vacuum, as a function of T, in the range 20–80 K, when the temperature of the second stage is 4.2 K. Figure 10.22 illustrates the case of liquefaction at 100 kPa and Fig. 10.23 the case of liquefaction at 210 kPa. In the figures, the integral power load lines of the first stage (IP_1) (red circles) and of the second stage ($\mathrm{IP}_2 + \mathrm{LP}$) (green triangles), the $\dot{m}/\dot{m}_0$ ratio (yellow circles) and the total heat power (purple circles) are also drawn.

Figure 10.22 nicely illustrates that, under perfect heat exchange between gas and cryo-refrigerator, the point of interception (gray dot) between the first stage load line ($\mathrm{IP}_1(T)$) and the load map power line of the first stage ($P_1(T)$) gives graphically $T_1 \approx 37$ K (vertical gray dashed line). Instead, as reflected on Table 10.1, the experimental value of T_1 is in this case smaller, 35 K (see vertical red dashed line), indicating that some extra heat (15%) is been extracted from the first stage regenerator at this flow rate. Typical values of extra heat power that can be extracted from the first stage regenerators are low, i.e., of the order of 1–20%, because the typical first stage regenerator materials are almost ideal and, consequently, their heat losses are low. Instead, the extra heat power that can be extracted from the much more non-ideal second stage regenerators can be very high. Second stage rare earth regenerators can accept much heat all along its length [22]. In this case $5.1 - 1.5\,\mathrm{W} = 3.6\,\mathrm{W}$!

Figure 10.23 presents the case for liquefaction at 210 kPa for which a much higher liquefaction rate of 34 L/Day (0.041 g/s) is obtained. The point of interception (gray dot) between the first stage load line ($\mathrm{IP}_1(T)$) and the characteristic load map power line ($P_1(T)$) gives in this case $T_1 \approx 53$ K (see vertical gray dashed line). Thus, contrary to the 100 kPa case, the experimental value of T_1 is now

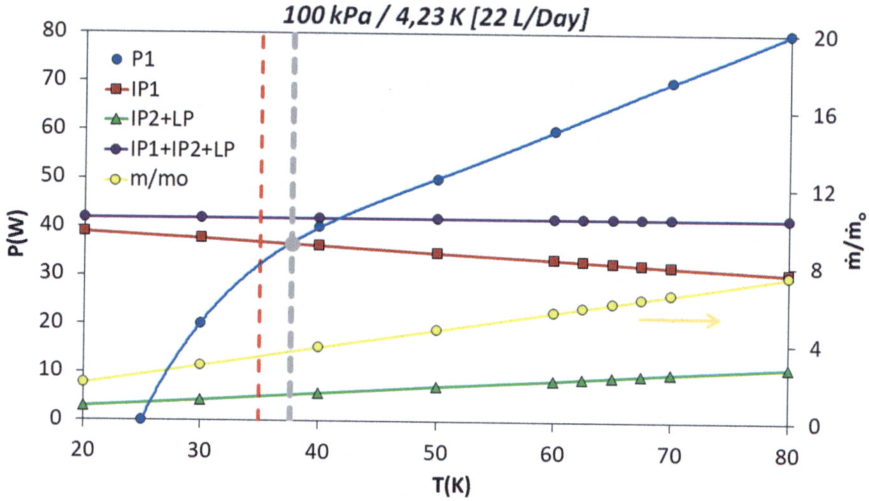

Fig. 10.22 Blue circles: first stage power $P_1(W)$ of cryo-refrigerator RDK-415 operating at 50 Hz, in vacuum, as a function of T, obtained from the load map in vacuum with $T_2 = 4.2$ K. Red squares: integrated power $\mathrm{IP}_1(T)$ required on first stage to precool 0.0268 g/s (equivalent to a liquefaction rate of 22 L/Day) down to $T_1 = T$. Green triangles: Integrated power IP_2 (T) plus liquefaction power LP required on second stage to precool and liquefy 0.0268 g/s from $T_1 = T$ to 4.2 K. Purple circles: total power required to precool and liquefy 0.0268 g/s from RT down to 4.2 K. The parameter $\dot{m}/\dot{m}_0$ is the ratio of the actual mass flow rate being cooled and liquefied to the mass flow rate that can be cooled and liquefied using only the 1.5 W cold station heat power at 4.2 K. The red vertical line indicates the measured value of T_1 for 0.0268 g/s. The gray vertical line corresponds to the expected T_1 for $\mathrm{IP}_1 = \mathrm{P}_1$

higher, 55 K (see red vertical dashed line), therefore $\mathrm{IP}_1(T = 55\text{ K})$ is smaller than $P_1(T = 55\text{ K})$, which means that part of the heat that should be available at the first stage cold station is now lost (about 10%). Thus, at the higher input gas flow rate of this case, imperfect heat exchange of the gas with the first stage is limiting the liquefaction rate.

Fortunately, the total heat power that is extracted from the non-ideal second stage regenerator and from the second stage cold station is as high as 12.05 W. Since the warm end of the second stage regenerator is now at a much higher temperature (55 K at 210 kPa vs 35 K at 100 kPa) the regenerator can accept much more heat all along its length [22]. In fact, in this case, the parameter that measures this effect, $\dot{m}/\dot{m}_0$, yields a value of 8.

10.5 Summary

Novel cryocooler-based techniques for small-scale helium liquefaction and purification have been developed. By applying a new approach to the traditional heat exchange methods, we have been able to double the previously reported

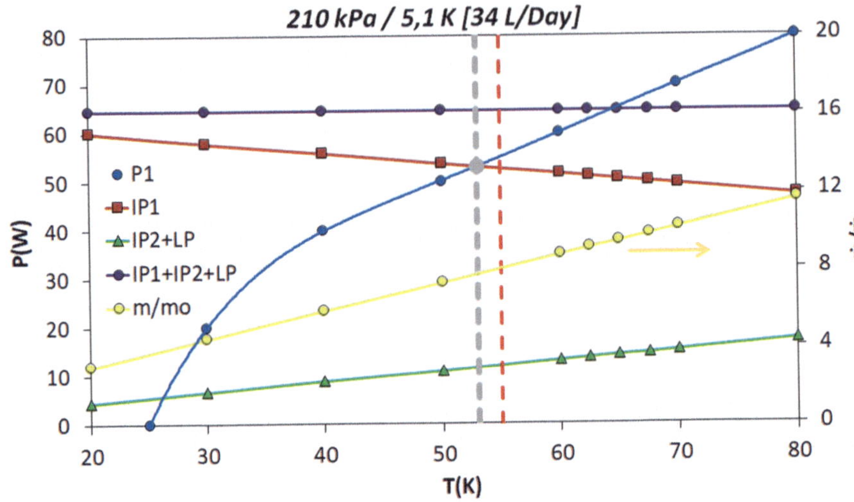

Fig. 10.23 Blue circles: first stage power P_1(W) of cryo-refrigerator RDK-415 operating at 50 Hz, in vacuum, as a function of T, obtained from the load map in vacuum with $T_2 = 4.2$ K. Red squares: integrated power $IP_1(T)$ required on first stage to precool 0.041 g/s (equivalent to a liquefaction rate of 34 L/Day) down to $T_1 = T$. Green triangles: integrated power IP_2 (T) plus liquefaction power LP required on second stage to precool and liquefy 0.041 g/s from $T_1 = T$ to 4.2 K. Purple circles: total power required to precool and liquefy 0.041 g/s from RT down to 4.2 K. The parameter $\dot{m}/\dot{m}_0$ (yellow circles) is the ratio of the actual mass flow rate being cooled and liquefied to the mass flow rate that can be cooled and liquefied using only the 1.5 W cold station heat power at 4.2 K. The red vertical line indicates the measured value of T_1 for 0.041 g/s. The gray vertical line corresponds to the expected T_1 for $IP_1 = P_1$

maximum liquefaction rates. The method uses elevated pressure to densify and finally condense the vapor at higher temperatures, taking advantage of the larger cooling power available from the cryo-refrigerator stages under these conditions, and, of the faster decrease of the helium enthalpy during the liquefaction process. Using cryo-refrigeration as a first step, and, chemical adsorption by non-evaporable getters, as a second step, to eliminate impurities in helium gas, we have been able to purify helium down to unprecedented purity levels. The "clean helium" gas thus obtained not only guarantees long-term performance and efficiency of small-scale liquefiers, but also allowed us to solve a world-wide contamination issue in standard evaporation cryostats: the blockage of low temperature impedances employed to reach temperatures below 4.2 K. From a theoretical standpoint, using a simple analytical model of mass and heat exchange balance, and, if heat can be removed all along the regenerator length and not only from the cold cooling station at the expansion volume of each stage, the experimental liquefaction rate values at different pressures have been interpreted. Heat exchange efficiency reduction for high flow rates counteracts the positive pressure effect on the liquefaction rate. Thus, the system and method so far developed have a liquefaction rate limit that is only obtained when an optimum liquefaction trajectory, that passes near or above the critical point, is selected.

References

1. Abrikosova II, Shalnikov AI (1971) The purity of liquid helium. Cryogenics (Guildf) 11:137–138
2. Barron RF (1985) Separation and purification systems. Cryogenic systems. Clarendon Press, Gloucestershire, pp 223-226
3. Chialvo C, Sayles T, Diederichs J, Spagna S, Gabal M, Sesé J, et al (2014) Smart energy compressors for helium liquefiers. In: Miller SD, Ross RG Jr (eds) Cryocoolers, vol 18. International Cryocooler Conference, Inc., Boulder, CO, pp 551–556
4. Collins SC (1966) Helium refrigerator and liquefier. Adv Cryog Eng 11:11–15
5. Cryomech. PT415. http://www.cryomech.com/cryorefrigerators/pulse-tube/pt415/. Accessed 31 Oct 2018
6. de Waele ATAM (2011) Basic operation of cryocoolers and related thermal machines. J Low Temp Phys 164:179–236. https://doi.org/10.1007/s10909-011-0373-x
7. Diederichs J, Simmonds MB, Sayles T, Spagna S (2012) A modular architecture for helium compressors larger than 2.5 kW. Cryocoolers 17:367-371
8. Gabal M (2017) New cryocooler-based helium liquefaction and purification techniques. From recovered gas to ultra-pure liquid. PhD thesis, Prensas de la Universidad de Zaragoza
9. Gabal M, Arauzo A, Camón A, Castrillo M, Guerrero E, Lozano MP, et al (2016) Hydrogen-free liquid-helium recovery plants: the solution for low-temperature flow impedance blocking. Phys Rev Appl 6:024017. https://doi.org/10.1103/PhysRevApplied.6.024017
10. Gabal M, Sesé J, Rillo C, Spagna S (2018) "Clean" liquid helium. In: Zivieri R (ed) Superfluids and superconductors. IntechOpen, pp 67–86. https://www.intechopen.com/books/superfluids-and-superconductors/-clean-liquid-helium
11. Gifford WE, Longsworth RC (1964) Pulse-tube refrigeration. J Eng Ind 86:264. https://doi.org/10.1115/1.3670530
12. Gschneidner K, Pecharsky A, Pecharsky V (2002) Low temperature cryocooler regenerator materials. Cryocoolers 12:457-465
13. Kuriyama T, Takahashi M, Nakagome H, Eda T, Seshake H, Hashimoto T et al (1992) Helium liquefaction by a two-stage Gifford-McMahon-cycle refrigerator using new regenerator material of Er_3Ni. Jpn J Appl Phys 31:L1206–1208. https://doi.org/10.1143/JJAP.31.L1206
14. Lemmon EW, Huber ML, McLinden MO (2013) NIST standard reference database 23: reference fluid thermodynamic and transport properties, version 9.1. The National Institute of Standards and Technology, Gaithersburg
15. McMahon HO, Gifford WE (1960) A new low-temperature gas expansion cycle. Adv Cryog Eng 5:354-367. https://doi.org/10.1007/978-1-4757-0537-9_44
16. Mikulin EI, Tarasov AA, Shkrebyonock MP (1984) Low-temperature expansion pulse tubes. In: Fast RW (ed) Advances in Cryogenic Engineering SE - 72, vol 29, Springer, New York, pp 629-637. https://doi.org/10.1007/978-1-4613-9865-3_72
17. Nagao M, Inaguchi T, Yoshimura H, Yamada T, Iwamoto M (1990) Helium liquefaction by a Gifford-Mcmahon cycle cryocooler, pp 1251–1260. https://doi.org/10.1007/978-1-4613-0639-9_149
18. Quantum Design. Model ATL160.
19. Quantum Design. Model ATL80.
20. Quantum Design. Model ATL160+.
21. Quantum Design. Model ATP30.
22. Radebaugh R, Marquardt ED, Gary J, OGallagher A (2001). Regenerator behavior with heat input or removal at intermediate temperatures. In: Cryocoolers, vol 11. Kluwer Academic Publishers, Boston, pp 409-418. https://doi.org/10.1007/0-306-47112-4_52
23. Rillo C, Gabal M, Lozano MP, Sesé J, Spagna S, Diederichs J et al (2015) Enhancement of the liquefaction rate in small-scale helium liquefiers working near and above the critical point. Phys Rev Appl 3:051001. https://doi.org/10.1103/PhysRevApplied.3.051001

24. Rillo C, Sesé J, Gabal M, Lozano MP, Téllez P, Campillo V et al (2015) Efficient small-scale helium liquefaction: from super-critical gas to super-cooled single-phase He I. IEEE/CSC ESAS Supercond NEWS FORUM, pp 6-11
25. Satoh T, Onishi A, Li R, Asami H, Kanazawa Y (1996) Development of 1.5 W 4 K G-M cryocooler with magnetic regenerator material. In: Kittel P (ed) Advances in Cryogenic Engineering SE - 205, vol 41. Springer, New York, pp 1631–1637. https://doi.org/10.1007/978-1-4613-0373-2_205
26. Schmidt-Wellenburg P, Zimmer O (2006) Helium liquefaction with a commercial 4 K Gifford-McMahon cryocooler. Cryogenics (Guildf) 46:799–803. https://doi.org/10.1016/j.cryogenics.2006.08.004
27. SHI Cryogenics RDK-415D 4K. http://www.shicryogenics.com/products/4k-cryocoolers/rdk-415d-4k-cryocooler-series/. Accessed 31 Oct 2018
28. SHI Cryogenics RDK-408D 4K. http://www.shicryogenics.com/products/4k-cryocoolers/rdk-408d2-4k-cryocooler-series/. Accessed 31 Oct 2018
29. Thummes G, Wang C, Heiden C (1998) Small scale 4He liquefaction using a two-stage 4 K pulse tube cooler. Cryogenics (Guildf) 38:337–342. https://doi.org/10.1016/S0011-2275(97)00169-0
30. Wang C (2009) Small scale helium liquefaction systems. J Phys Conf Ser 150:012053. https://doi.org/10.1088/1742-6596/150/1/012053
31. Wang C (2001a) Helium liquefaction with a 4 K pulse tube cryocooler. Cryogenics (Guildf) 41:491–496. https://doi.org/10.1016/S0011-2275(01)00116-3
32. Xu M, Takayama H, Nakano K (2008) Development of high efficiency 4 K two-stage pulse tube cryocooler. Cryocoolers, vol 16. Georgia Institute of Technology, Atlanta, p 57
33. Xu MY, Gao JL, Seitz E, Longsworth RC, Saito M, Ikeya Y (2006) Development of SHI 1.0 W 4 K two-stage pulse tube cryocoolers. In: AIP conference proceedings, vol 823. AIP, College Park, pp 837-844. https://doi.org/10.1063/1.2202493

The manufacturer's authorised representative in the EU is Springer Nature Customer Service Centre GmbH, Europaplatz 3, 69115 Heidelberg, Germany. If you have any concerns regarding our products, please contact ProductSafety@springernature.com

Printed and bound by CPI Group (UK) Ltd, Croydon, CR0 4YY
22/09/2026
02221062-0009